THE
SCIENTIFIC APPROACH

*Basic Principles of the
Scientific Method*

THE
SCIENTIFIC APPROACH

Basic Principles of the Scientific Method

by

CARLO L. LASTRUCCI

San Francisco State College

SCHENKMAN PUBLISHING COMPANY, INC.
CAMBRIDGE, MASSACHUSETTS

Copyright © 1963, 1967

SCHENKMAN PUBLISHING COMPANY, INC.
Cambridge, Massachusetts 02138

Third Printing

PRINTED IN THE UNITED STATES OF AMERICA

Foreword

Man is now witnessing the second great ideological onslaught of science, and he is both awed and appalled by what he sees. The first great turmoil in Western thinking occurred in the Middle Ages when the nascent science of Copernicus, Galileo and Newton forced medieval man to re-appraise himself in relation to the cosmos. Today, however, science is forcing man to reorient himself in relation to his potential survival on his own shrinking planet. To say that science now holds both the key to a better life and a trigger to complete annihilation is not metaphorical whimsy; it is all too literally true. For if modern man cannot devise ways to control this staggering potential for life or death within the near future, the probability of his very survival will become a question of serious doubt.

The time undoubtedly has arrived when every thoughtful person should have an understanding of both the distinct limitations and of the reasonable potentialities inherent in scientific method. No one can any longer afford to remain either ignorant of, or craven toward, this most potent force in man's history. To remain so is to invite exploitation at least, or destruction at worst. For ignorance of science or servility toward its use encourages unprincipled individuals to employ it for selfish or even destructive ends—whether in the seductively distorted blandishments of political slogans, on the one hand, or in the deadly nose cone of an atomic missile, on the other.

In short, both everyday necessity and ultimate survival demand that every thoughtful person understand what science actually is: namely, an objective and reliable intellectual enterprise, a method of systematic analysis of phenomena, a logical form of problem solving, which in itself is neither—and yet potentially can be made either—good or evil. The layman should understand why quackery is often mistaken for science and therefore accepted just as blindly; why some scientists at times speak either unintelligibly or erroneously, or even both; and why science in a free society can no longer be regarded as the esoteric bailiwick of an atypical band of devotees. In short, if the citizenry of the so-called free world are to influence constructively their moral, political, economic and physical destinies in which science will play an ever-increasing role, they will need to understand both what science is and is not, and what it can and cannot do.

This volume attempts to outline the essential features of the scientific approach. It attempts to synthesize and coordinate in nontechnical form the essential elements of two major fields of study: the logic and philosophy, and the methods, of science. It is not, however, intended to serve as a substitute for a scholarly treatment of those fields. Rather, it has been written for the potential reader who is seriously concerned about the logical structure and method of science but who does not have the time or the background to pursue these fields intensively.

The style perhaps demands a word of explanation. The writer has attempted to lower (if not actually eliminate) the semantic and conceptual barrier which often separates the lay reader from the literature of science. The absence of footnotes, for example, is purposive; and is justified on two counts: (1) The materials herein presented are accepted as a matter of fact well established in the common domain of scholarship, and therefore do not require substantiation by reference to particular authoritative sources. For this reason, (2) parenthetical remarks or definitions are included within the appropriate portions of the text in order to promote unhampered reading and clear understanding.

In spite of the conscientious attempt to employ words precisely, this objective has not always been achieved in the pages to follow. The literature of science is not yet highly standardized. The same term is often used in different ways by various writers or occasionally even by the same writer. The terms *validity* and *reliability*, for example, mean one thing to a logician but another to a statistician. Among the more common examples of key terms in the literature of science which are employed in a variety of ways are: *valid, true, reliable, verify, hypothesis, theory, law, principle, axiom, postulate, proof.* Whenever demanded for clear understanding, therefore, the specific (either denotative or connotative) meaning of such terms has been indicated.

The annotated bibliography has been purposely limited to a functional selection of sources which—when studied in whole or in large part—would provide adequate treatment of all the basic materials covered in the text. Many of these sources also contain very extensive bibliographies of their own, thus representing much of the best literature available on the many aspects of scientific method. It should be mentioned, however, that the plethora of books increasingly available which are somehow related to science includes many treatments which are either outmoded or seriously limited in both breadth and depth. At this present writing no single book exists which comprehensively or definitively covers all the major materials necessary to a full understanding of science; and it is highly doubtful that such a book could ever be written within the confines of a single volume.

As the text indicates throughout, this writer stresses the unity of science in all its branches. The discussions relate to science as a whole in terms of its common attributes. Illustrations, for example, purposely have been selected from the four major areas of scientific endeavor: the physical, the biological, the psychological and the social. One exception to this general approach, however, has been made in the interest of utility to the reader, to wit: The average person goes through life never having contact with a linear accelerator, an electron microscope, or any of the other complex instruments employed in the physical or biological sciences. But he can hardly avoid being involved with polls, questionnaires, rating scales, histories, and the other common instruments employed in the behavioral sciences. Thus, because of common experience and practical utility, special attention has been given in a few places to the particular problems involved in the collection and interpretation of social science data. Such attention, however, has been kept to a minimum.

The writer wishes to express his appreciation for all the assistance given either directly or indirectly by his many colleagues, former students and instructors, and the host of half-remembered writers whose works have in some known (and undoubtedly many unknown) ways contributed to the ideas presented in this volume. The indebtedness to all these contributors is manifold; but specific gratitude is expressed to my present colleagues and friends: Drs. D. L. Adler, D. L. Garrity, D. C. Gibbons, J. W. Kinch, R. T. LaPiere, R. A. Thornton, and R. Weingartner. Responsibility for the presentation, however, remains solely with the writer. It is hoped that this small volume will repay all the efforts of its many contributors by fostering an increased understanding and appreciation of the great contributions which science has made and can continue to make to the solution of man's many problems.

San Francisco, California

Contents

PART III
SCIENTIFIC ANALYSIS

Contents

Figures

Part I

THE
SCIENTIFIC ORIENTATION

CHAPTER I

What Science Is

A. DEFINITION

A. Definition

I.1 In spite of the tremendous influence of science upon modern civilization, there exists as yet no standardized definition of science. Laymen, scholars, and scientists themselves define the term in varying ways and employ it in a variety of contexts. To some, science simply denotes a generalized prestige (e.g., Christian Science, the science of boxing, scientific history, a scientific analysis of modern art). To others, it denotes a body of verified knowledge (e.g., biology, astronomy, physics, chemistry). To still others it connotes an objective analysis of phenomena. According to a standard dictionary, the word "science" is derived from the Latin word *scientia* (*sciens,* the present participle of *scire,* to "know"); and is defined in a variety of ways: (1) originally, state or fact of knowing; knowledge, often as opposed to *intuition, belief,* etc. (2) systematized knowledge derived from observation, study, and experimentation carried on in order to determine the nature or principles of what is being studied. (3) a branch of knowledge or study, especially one concerned with establishing and systematizing facts, principles, and methods, as by experiments and hypotheses: as the *science* of music. (4) *a.* the systematized knowledge of nature and the physical world. *b.* any branch of this. (5) skill, technique, or ability based upon training, discipline, and experience: often somewhat humorous, as the *science* of boxing.

I.2 Checking the definition of the adjective "scientific" does not help very much to determine its essential features. According to a standard source, the word "scientific" is derived from the Latin word *scientia,* meaning knowledge, plus the term *facere,* meaning to make; both terms were originally employed as a translation from the Greek term *episthemonikos,* or making knowledge (from which the modern term *epistemology* is derived, meaning the study or theory of the origin, essence, methods, and limits of knowledge). In German, for example, the term for science, *Wissenschaft,* denotes these same two qualities: *wissen,* to know, and *schaft,* to do, make, or work. This term, like the Greek form, connotes a special feature of this type of knowledge: namely, applied knowledge or applied logic (from the Greek term *logos,* meaning to know). In short, the word "science" connotes reasoned knowledge and applied logic.

I.3 The difficulty encountered in attempting to define the term "science" arises mainly from the tendency to confuse the content of science with its method. Much of the content of science is constantly changing; so what may be scientific (i.e., accepted as true) today may become unscientific (i.e., regarded as untrue) tomorrow. (A common illustration of this point is the case of *phlogiston,* which reputable physicists once thought was the necessary agent for combustion to occur. Now everyone knows that oxygen is the necessary agent in combustion, and *phlogiston* does not exist in modern scientific thinking.) Furthermore, the demarcation between science and nonscience is not clear-cut; it actually is not a line but an area both shifting and subject to debate—depending upon the authorities one accepts. For purposes of precision, clarification and usefulness, therefore, it seems appropriate to define the term "science" in a combination of three ways: (a) analytically—i.e., in terms of its essential and distinctive component attributes, (b) functionally—i.e., in terms of the services it performs, and (c) operationally—i.e., in terms of the processes or operations performed when practicing it.

I.4   An examination of scores of standard books about science fails to elicit a clear and comprehensive definition; but such an examination does rather clearly suggest a consensus among authoritative writers with regard to the essential attributes or processes of science. According to such a consensus, *science may be defined* quite accurately and functionally as: *an objective, logical, and systematic method of analysis of phenomena, devised to permit the accumulation of reliable knowledge.* The following discussion will be devoted to an explication of the key terms of this definition, while this book as a whole attempts to serve as an operational definition of the scientific method.

I.5   *Objective:* Science is an intersubjective method; it is available to any interested and competent person; it is not the special province of a favored few. Objectivity in science refers to attitudes devoid of personal whim, bias or prejudice, and to methods centered around ascertainment of the publicly demonstrable qualities of a phenomenon. Evidence in science is factual, not conjectural; and truth is achieved by the demonstration of evidential proof. Though science is a subjective enterprise insofar as it is practiced by individuals, scientific method encourages a rigorous, impersonal mode of procedure dictated by the demands of logic and objective procedures. Authority in science is achieved by the accumulation of publicly ascertainable evidence supporting one's argument; it is not a consequence of mere opinion (no matter how strongly held), nor of faith alone, nor of assumed verity. To be devoted to the ideal of objectivity in science is, in effect, to be devoted to the ideal of public verifiability according to the consensus of trained observers.

I.6 Objectivity also connotes an attitude devoid of subjective value judgments. The attitude of the scientist toward *normative* (i.e., value-endowed, expressing a desirable norm or standard) phenomena is treated later; but suffice it to say at this point that objectivity in science denotes an impartial viewpoint which refrains from inferring or implying that any phenomenon is "good" or "bad" *per se*. Whether or not absolute objectivity is ever possible for human beings to achieve is a question best left to philosophers; but certainly the scientist attempts constantly—both by training and particularly by the use of objectifying instruments —to look at his data with as little bias as possible.

I.7 The objective attribute of science specifies that the ultimate court of resort in any speculative argument is the objective event: an observation or any experience that can be publicly verified by trained observers. Pending a detailed discussion later of how data in science are derived, it should suffice to point out at this time that the impressive growth of modern science as a reliable method of achieving knowledge is based largely upon its foundation of verified objective facts. To the extent that a particular field of study employs non-objective data or refers to subjective phenomena, then to that extent is it regarded as "less scientific" (i.e., less reliable) than are the more advanced sciences.

I.8 To say that science is objective also connotes that it is *descriptive* and *analytical*. In practice scientific method simply answers the question of "What is the fact?" or "What is the relationship?" When scientists sometimes advocate or prescribe, they do so only to answer a legitimate scientific question, and are saying in effect *"If* this, . . . *then* this." Although scientific method raises many questions of its own, they are always of a logical, of a methodological, or of a meta-ethical (i.e., universally ethical) order; they are never of a parochially ethical order. That function belongs to religion, politics, or other normative systems of belief; it can never be imposed upon, delegated to, or assumed by science.

I.9 To anticipate somewhat a more detailed discussion of this point, it might be said here that this attribute probably occasions more misunderstanding among laymen than does any other attribute of science. True, scientists offer many answers to normative questions, but such questions are generally not in themselves of scientific origin. A scientist might say, for example, "You should fly above twenty-thousand feet, *provided* you want to conserve fuel"; or "You should take insulin, *provided* you want to avoid death from diabetic shock." He likewise might say "We should adopt a city-manager plan, *provided* it has been demonstrated to be a more efficient type of government, and *assuming* that efficiency in government is desirable." In short, the scientist is never prescriptive or advocatory except (a) when he is functioning simply as a layman and not as a scientist as such, or (b) when he has been addressed with a value question concerning which he possesses objective knowledge.

I.10 The foregoing remarks should not be interpreted to imply that a scientist, as a scientist, does not advocate any values. He does, and he holds such values highly. He values, for example, the pursuit of knowledge for its own sake (i.e., simply for the satisfaction of knowing); he values the attributes of science herein being discussed as desirable qualities in relation to his work; and he values the opportunities to pioneer on the frontiers of research. In other words he values highly what may be referred to as the spirit of inquiry and the ideal of objectivity (or "truth"). These values underlie all his efforts as a scientist, but they may be and often are quite distinct from the popular values of his culture as such. As a scientist, about the only way that he is influenced by cultural values would be in his choice of problems for research, and of course in the impetus which led him to become a seeker of knowledge in the first place. In this sense of the term "value," it is undoubtedly true that every person, including the scientist, is a product of his time and culture, and therefore cannot escape the influences of its value system.

I.11 A final note on the subject of objectivity concerns the question of the so-called normative sciences. During its present evolution modern science has developed largely in the areas of physical phenomena. But there is no inherent reason why the scientific approach cannot be applied to areas involving human attitudes, beliefs and values. In fact modern scientific psychology shows exactly this tendency. At the present time the objective approach to the analysis and understanding of subjective or mental phenomena has been beset with staggering methodological difficulties; but there is no inherent reason to assume that both the theoretical and technical problems involved in studying subjective behavior cannot be solved in the future. In time, when the social and psychological sciences have achieved greater knowledge, rigor and skill, a true science of subjective phenomena (e.g., ethical, moral, artistic feelings) is certainly forseeable.

I.12 *Logical:* To say that science is a logical method is simply (yet very significantly) to say that the scientist is constantly guided by the accepted rules of reasoning standardized by reputable logicians. Competence in science demands competence in logical analysis—a large field in itself, and some crucial aspects of which are noted throughout this book. Rules of definition, forms of deductive inference, theories of probability, systems of calculi, etc., are fundamental in any mature and reputable science. Science is not a "cook book" of facts, rules or formula. Rather, it is a *systematic* arrangement of facts, theories, instruments and processes —all interrelated by principles of reasoned thought. Although one may function in applied fields (e.g., dentistry, plumbing, auto repairing, bridge building) by learning and applying formulas, to function as a scientist demands both a thorough grounding in logical analysis as well as skill and knowledge of a specific factual sort.

I.13 Since science is a logical system, it is also self-critical—i.e., it contains within its methods the tools for its own analysis. Somewhat like a physician checking his own health, the scientist

in practice constantly employs skepticism toward current as well
as new ideas, constantly retests accepted facts, constantly suggests
doubt when answers appear too obvious, and constantly demands
that all new ideas be subjected to the merciless demonstration
of objective verification. As a safeguard against intellectual smug-
ness or dogmatism, this attitude on the part of the scientist be-
speaks a fundamental awareness of the fact that the desire to
believe can be intellectually seductive even for the best minds.
So in order to avoid the all-too-human tendency to be satisfied
with the *status quo,* the scientist guards himself by a cloak of
critical doubt. In this sense, science can never offer the comfor-
table surety of omniscient systems of belief. The true scientist
is constantly searching, never satisfied, and always doubtful of
everything he knows, for he has learned early in his career the
singular lesson of science: what is thought to be true today might
be proved to be false by tomorrow.

I.14 *Systematic:* Science proceeds in an orderly manner both
in its organization of a problem and in its methods of operation.
It does not proceed randomly or haphazardly. Here is one of the
qualitative differences between scientific and popular thinking.
Unscientific analyses tend to marshall sundry and often unrelated
facts in order to support an argument; violating, at the same
time, the accepted principles of logical inference in support of
their proofs. The systematic procedures inherent in the scientific
approach assume the form of a closely interwoven, logically arranged sequence of steps which permit little deviation. Verification in science, as a matter of fact, is substantially a systematic
process of logical inference which demands that the premises,
facts and conclusions be arranged in an orderly manner.

I.15 The systematic feature of science also implies internal
consistency. In a well-developed science, the various theories and
laws are interrelated and corroborative. They support each other,
or at least do not contradict each other. Immaturity of a science
is characterized by serious internal disagreements relating to
theories, laws, propositions, principles and sometimes even of
method. It should be pointed out, however, that complete and
final internal consistency is never achieved even in the most ad-

vanced sciences. New discoveries suggest new laws, principles or theories, which in turn demand the modification of established notions of reality. It is in this way that science grows. This dynamic quality, however, (except in a beginning science) should not be construed to permit glaring contradictions in the internal interrelation and interdependence of the theoretical foundations of that science.

I.16 *Method:* Science is a systematized form of analysis; it is not any particular (or even general) body of knowledge. But a dispute often arises as to whether there is only one scientific method or several. Pending a detailed discussion of this point in Chapter V, it may be said at this time that this problem is largely a semantic one, not a conceptual one (i.e., it arises mainly from the various meanings attached to the word "method"). This writer takes the position that there is basically *only one* scientific method, however broad its borders may be; and that the various so-called "methods" are essentially techniques of research or exemplifications of the basic general method of description and analysis as defined in the previous discussion and as developed throughout this text.

I.17 Though the various fields of science differ both in content and in techniques, an examination of all the highly developed sciences reveals a common foundation of procedures of inquiry. It is in this sense—of possessing common essential features —that the singular term "method" is here preferred to the plural term "methods." To contend that different fields of study utilizing basically different methods of inquiry are all equally valid (hence, "scientific"), is to suggest that method is not significantly related to validity. Yet all reputable scientists follow the same basic rules of evidence and reasoning in order to validate their conclusions. It is difficult, therefore, to accept the notion that basically different methods of inquiry—not simply different techniques—are all equally valid, therefore equally scientific. It is hoped that this volume will help not only to demonstrate the essential features which give scientific method its unity, but also will help to provide a functional and meaningful description of scientific method at its best.

I.18 *Phenomena:* Scientific method is applicable to any kind of behavior or event that has objectively demonstrable attributes or consequences. If an event is presumed to be inherently subjective (e.g., an idea, a feeling, an inspiration, a dream), then it is not amenable to scientific analysis— unless, of course, its presence can be demonstrated by virtue of some related objective attributes or consequences. Thus, for example, a presumably subjective behavior (say, a dream) cannot be studied scientifically until and unless it can be shown to exhibit objective attributes or consequences. Though the phenomena studied by science are publicly verifiable, it should not be assumed that such objects or events of study are the only interest of science. Scientific method is built upon a foundation of ideational abstractions (i.e., notions, ideas, theories, laws, principles, etc.) devised to relate and explain observable objects and events. Much of the content of science, therefore, consists of intellectual notions about things or events. But the object of all such thought is the particular phenomenon being studied; and that particular phenomenon is the experience or perceived event.

I.19 A discussion of the legitimate object of scientific study (i.e., the observed object or event) often raises the question of assumedly spiritual (i.e., extra-natural) or supernatural phenomena. Put another way, it is often asserted that science cannot study—and therefore cannot legitimately answer questions about —supernatural events. The scientific answer to such an assertion is quite simple. Philosophically, science is neither theistic (i.e., a belief in a god or gods) nor atheistic; but it is objective insofar as it studies phenomena having behavioral attributes or consequences. If something exists or occurs having such attributes that it can be objectively ascertained and confirmed, then it can be studied scientifically. Whether it has been "caused" by a "natural" or a "supernatural" force or agent is a separate and wholly different kind of question (which will be considered later in some detail).

I.20 *Devised:* It is often remarked that science is a "synthetic" system insofar as it is an artificially created synthesis of various elements into an interrelated and logical whole. In short, it is an invention, a creation of human ingenuity. The only remark that need be made about the term "devised" is that scientific method is a creation to serve a particular purpose—viz., the orderly arrangement of factual knowledge and ideas about reality in that form which seems most fruitful in terms of the end to be served. Since it is created to serve a particular purpose, it can—and in fact in some ways, does—change as new ideas suggest modifications. But the only essential point that needs to be made here is that man arranges his thinking about his world according to various preferences; and the scientific method is the one such arrangement which has so far appeared to be the most fruitful for the explanation of objective phenomena.

I.21 *Accumulate:* Science is cumulative; it is an integrated system built up in an orderly manner wherein each fact, principle, theory, law, etc., supports other facts, theories, laws, etc. But science is not a mere accumulation. A cook book, a telephone directory, a stock-market report, an engineer's manual, are all assemblages of accumulated facts—but they are not science. The accumulated knowledge making up a systematic science is dynamic, not static. Science always seeks additional knowledge in the belief (borne out by history) that knowledge is never complete. "Truth" in science is always relative and temporal, never absolute or final. In contrast to many closed philosophical or ideological "systems" (e.g., political, aesthetic, moral, religious, and other theories with which we are familiar), science may properly be defined as an "open" rather than as a "closed" system of ideas. Therefore, it grows constantly by discarding erroneous or useless notions and by substituting more correct or useful ones in the light of new evidence. As a matter of fact, this very dynamic quality is one of the intrinsic features of science; and primarily because of this, science has grown continually from its primitive beginnings to its present awe-inspiring eminence.

I.22 This attribute of cumulativeness certainly should not be construed to mean that science grows by simple accretion. Scientific theorists employ a related *principle of parsimony* (often referred to as Occam's razor) which decrees that complex explanations or methods should be replaced by simpler formulations wherever possible. In fact, the tendency of scientists to designate some phenomena by "big words" actually is a consequence of this principle; for the history of science demonstrates that previously complex explanations or designations are constantly being replaced by scientifically simpler and more precise terminology. (This principle in action is illustrated in several different sections throughout this book.) This principle suggests, for example, that when there seems to be equal probability of truth, utility or fruitfulness among various techniques or theories, the simplest one should be preferred. Simplicity, however, does not necessarily connote ease of comprehension. As employed in this principle, simplicity refers to a preference for the smallest necessary number of independent elements to be embodied in a theory or a procedure.

I.23 More important, the principle of parsimony decrees that one should strive constantly to explain as much as possible by the employment of as few terms, symbols, concepts or formulae as possible. In a broad sense, then, a major function of science is to explain all its phenomena as parsimoniously (i.e., as economically) as possible. An outstanding virtue of Newton's laws of motion, for example, was the fact that not only did they explain many formerly unexplained phenomena, but that they also explained the laws of falling bodies, "discovered" by Galileo earlier, as well as the movements of heavenly bodies, described by Kepler earlier. Thus the attribute of cumulativeness and the principle of parsimony are so closely interrelated that science strives constantly to predict the behavior of as yet unobserved phenomena in terms of the common and known qualities which they possess as members of a class of phenomena. Cumulativeness, orderliness and parsimony, therefore, work together to permit the largest possible number of specific predictions to be made from as few basic and general "laws" as possible.

I.24 The term "reductionism" is often employed in discussions of this broad principle of parsimony. Essentially, reductionism refers to the general practice of striving to encompass as many sub-theories as possible into broader, more-inclusive categories of "grand theories." Though much scientific knowledge at any given time is temporarily unrelated or uncoordinated (e.g., discrete facts or "laws"), the scientist strives constantly to interrelate such discrete facts into meaningful wholes or patterns.

Eventually—as verified by the history of science—such patterns become integrated into larger systems of facts and ideas ("theories") which permit a wider range of explanation than would have been possible had the segmented facts been viewed in isolation.

I.25 *Reliable* (knowledge): This term has several and somewhat different specific meanings, but in the present context it refers to that kind of knowledge which one can depend upon in terms of predictability. In this sense, then, reliable knowledge is synonymous with exact or correct knowledge. Science strives constantly for exactness; it is not satisfied with half-truths and is intolerant of careless procedures. Probably the outstanding quality both in popular thinking and in the professional attitudes of scientists is this feature of precision and exactness. In fact it is almost axiomatic that science progresses to the extent that its measurements and calculations become more refined. It should be borne in mind, however, that accuracy or precision is not an end in itself but relative only to the purposes which it is to serve: viz., to promote more specific description, and hence to promote reliable prediction or control. Many exact measurements are meaningless in terms of the purpose at hand. There is at present no useful purpose served, for example, in refining the distance between the earth and the sun in terms of millimeters (if such could in fact be done). The gross measure in terms of thousands of miles is accurate enough for present scientific purposes.

I.26 Furthermore, many significant scientific principles are not supported by precise facts. The genetic principles involved in Mendel's laws of heredity, for example, do not yet permit a highly specific determination of the results of cross-breeding among large and heterogeneous populations. Nevertheless, increasing exactness usually generates increased understanding. Therefore, the highest degree of exactness commensurate with the demands made by the problem is an outstanding attribute of scientific method. As will be noted many times throughout this volume, a common stumbling block to the further development of a scientific approach is often the very inexactness of much present knowledge.

I.27   The problem of achieving reliability is a highly complex one, and will be examined in its various aspects throughout much of this book. At the present, however, reliable knowledge generally refers to knowledge which permits better predictions than could be made by chance or guesswork alone. Admittedly much popular ("folk") knowledge is reliable insofar as much common behavior is highly predictable by virtue of habit, custom, familiar experiences, etc. But much popular prediction, when accurate, is so simply by virtue of chance alone. The function of scientific method, therefore, is to understand phenomena to such an extent that the ratio and scope of accurate predictions can be consistently increased. It is presumably only through a valid, organized system of knowledge such as science that prediction can be effectively raised beyond the limited experience of a particular and unsophisticated group of individuals. Illustrations of this fact can, of course, easily be drawn from any common field where the knowledge and skill of the scientist is pitted against that of the layman.

I.28   The preceding discussion explicating the definition of science has left unanswered a basic question: Is scientific method the only reliable method—or is it simply one of many equally valid methods—for answering queries about objective phenomena? This question can be answered only in relative terms in light of one's faith in various knowledge systems. Among scientists, of course—and presumably among most laymen too—it is safe to aver that scientific method is regarded as the most reliable method so far devised for understanding objective phenomena. Such a statement implies, in effect, that in the competitive arena of opposing logical and methodological systems, all presumably devoted to the discovery of truth, scientific method has become the strongest intellectual tool that man has devised for furnishing verifiable and practicable answers to questions of demonstrable fact. Stated otherwise, this conviction avers that man has tried many different systems for answering the factual problems posed by himself and by nature; and that of all such systems, scientific method has unquestionably achieved the highest reliability of all.

I.29 The imposing growth of the layman's faith in scientific method has been derived from the ability of science to answer questions of a physical and biological nature. But the remarks in the previous paragraph should not be construed to be limited to nonhuman (and particularly to nonsocial) phenomena. This conviction of the efficacy of scientific method also implies that human problems of fact—as distinct from problems of faith and morals (i.e., of purely subjective value judgments or entirely normative questions)—can be solved better by the employment of a scientific approach than they can by other methods. This implies, for example, that many psychological and social problems which have perplexed man for centuries can eventually be solved satisfactorily provided (a) that they are submitted to a scientific approach, and (b) that they are phrased in objective terms. This problem will be dealt with more thoroughly in the following chapters.

I.30 A final remark relevant to the definition of science should be made concerning the proper scope of scientific inquiry. According to some views, scientific method can profitably be applied to any field of human endeavor; others argue that it is inherently limited to purely physical and biological phenomena of a nonsocial sort. In fact in some fields (particularly history and the humanities) there is still a wide range of opinion as to the feasibility of utilizing the scientific method at all. Perhaps this volume will help to clarify if not resolve this issue; for throughout various chapters it should become quite evident that it is *not the field* of study but the *type of problem* posed that determines whether or not a scientific approach can be profitably employed. At this point, therefore, the *scope* of scientific method will be viewed as the *whole range* of human interest.

B. *Science* versus *Pseudo-science*

I.31 Science as herein defined refers to a method of analysis involving (a) certain basic procedural attributes (objectivity, exactness, etc.), as well as (b) certain basic assumptions about reality (viz., the "postulates" of science to be explained in the following chapter). Pseudo-science, therefore, refers to modes of analysis which *pretend* or *profess* to meet the requirements of scientific method but which in fact violate one or more of its essential attributes. Flagrant examples of pseudo-science are easy to identify; but the more subtle—and therefore more insidious and convincing—cases require a rather precise delineation of the attributes involved.

I.32   Before examining pseudo-science, however, it is important to realize that knowledge is advancing rapidly. What may have been regarded as legitimate science at one time or place may later be viewed as pseudo-science. Phrenology, for example (i.e., "reading character" by interpreting the structure of the skull), was at one time regarded as a legitimate type of psychology. Today it has become simply a device for duping ignorant laymen. Astrology (i.e., predicting the future by the stars), palmistry (i.e., predicting a person's future by "reading the life lines" on his palm), or numerology (i.e., predicting one's future by interpreting the order of numbers in his birth date, or the numbered-order of the letters in his name, etc.), were all at one time regarded as reputable sciences. Today they are clearly defined as quackery.

I.33   A second aspect of this problem of identifying pseudo-science is a consequence of the fact that some methodological approaches are partly scientific and partly pseudo-scientific. That is, they combine legitimate with illegitimate methods, inferences or assumptions. Two cases in point might be cited: (a) religious "science"—which rests partly on objectively verified psychological principles of suggestibility (i.e., "mind over matter"), and partly on scientifically unconfirmed notions of bodily processes (i.e., all illness is "mental"); and (b) naturopathy— which rests partly on verified physiological principles (viz., the relation of diet to health), and partly on scientifically unconfirmed notions of bodily processes (viz., that all diseases can be cured without resort to medicines or surgery).

I.34   The essential features of a pseudo-scientific approach to phenomena, therefore, are those which ignore, deny or violate the essential attributes of valid science. Among the outstanding qualities of a pseudo-science are the following: (a) It is usually subjective rather than objective—i.e., it relies upon the unique personal interpretations of phenomena made by a particular practitioner. Therefore, it varies with the particular "authority" (e.g., a dogma, a bible or its equivalent, a semi-god, some type of oracle, or a charismatic leader—i.e., one perceived by his or her followers as having god-like qualities). (b) It is illogical insofar as it violates one or more basic rules of inference, of definition, of argument, of proof, etc. (c) It is unsystematic in the respect that its various parts do not necessarily relate to and support each other by virtue of internal consistency. (d) It is "fixed" or "closed" rather than accumulative insofar as its "facts" are unimpeachable in terms of new evidence. (e) Just as important is the fact that it exhibits very low reliability—i.e., its predictions are no better than those one could make by chance. Since an ultimate criterion of the validity of any purported science is its predictive power (a subject to be discussed at some length later), it is particularly in this respect that pseudo-science—when subjected to valid and objective measures of prediction—performs no better than does random guesswork alone.

I.35 Since the layman is untrained in the intellectual aspects of modern science, he often finds himself in a quandary when trying to distinguish between legitimate scientists and their imitators. All around him, every day, he hears or sees apparent authorities (often dressed in white laboratory coats) who exhort him to believe their assertions. The resolution of this quandary is an overwhelming task for the layman; for who and what is reputable and therefore reliable in science is a question demanding knowledge both of science in general and of the field involved in particular. It is for this reason that the nonspecialist is so often confused, and even duped, by conflicting claims all made in the name of science. This quandary, incidentally, faces the scientist himself at times when he is faced with cogent arguments outside his field of special knowledge.

I.36 Of course, many purported scientists are simply flagrant charlatans; but an additional difficulty facing the layman is that of differentiating the scientist when he is functioning as such from the scientist in his role as an ordinary citizen. Many reputable scientists speak or write publicly on many topics outside their established fields of accomplishment. A scientist—at least in

a so-called free society—has the same right as has any other
citizen to speak on any topic he wishes. But when a reputable
scientist purports to speak authoritatively *outside* his field of
knowledge, he is then exploiting the "halo prestige" of his scientific role (i.e., he is exploiting the tendency of individuals to
extend a person's prestige from one field to another but unrelated field of accomplishment). The physicist who speaks authoritatively in the field of politics, the psychologist who speaks
authoritatively in the field of child-rearing—all are as pseudo-
scientific at that time as is the opera star who endorses deodor-
ants, or as is the famous athlete who pontificates on the effects
of smoking. In these cases, such persons are functioning pseudo-
scientifically; and their established prestige in one field does not
qualify them to speak authoritatively in any field other than the
one of their established competence.

I.37   One particular facet of this science *versus* pseudo-science
problem which often seems to interest the layman is the role
of the amateur in relation to scientific discovery. The notion is
commonly held that great discoveries are often the product of
amateur minds, and therefore that the authority of the scientist
is sometimes to be critically suspected. This popular notion contains both a philosophical implication and an intellectual imputation. The philosophical implication is that discovery is
solely a matter of accident; and therefore that discovery in science
is essentially no different than, say, the discovery of a cache of
gold buried under the sands of a beach. With most great scientific
discoveries, however, the reverse is often true.

I.38   The intellectual imputation of this popular notion, however, is more complex. It appears obviously demonstrable that
the scientist is highly regarded by most laymen, sharing as he
does the generalized prestige of all intellectuals and professionally trained persons—especially those whose efforts so often result
in positive contributions to mankind's material welfare. In some
respects, in fact—especially since the significance of scientific
achievement both in the medical and in the astrophysical fields
has become highly magnified—one might safely generalize that
the scientist has become somewhat of a "hero type."

I.39 Yet at the same time, many people of limited knowledge seem to suspect or resent the superior abilities of prestiged individuals—especially so in the United States, where the doctrine of political equality is often extended to imply both social and intellectual equality. In this sense, then, the scientist—like the college professor, to whom he is professionally very closely related—is often regarded at best as a somewhat impractical and ineffectual person who spends most of his time on inconsequential matters. (This is the "absent-minded professor" of cartoon fame.) At worst, however, he is sometimes viewed by laymen as a sinister or even malevolent individual whose "cold-blooded" devotion to scientific discovery knows no bounds of humanistic or ethical values. (This is the "mad scientist" of horror-movie fame.) The social significance of this dualistic attitude of laymen toward the scientist invites serious speculation, but is beyond the scope of this volume.

I.40 Regardless of its philosophical implications or its intellectual imputations, this popular notion in regard to discovery is simply erroneous in fact. True, a few "great" discoveries that later were scientifically encompassed or explained have been made by poorly trained workers or laymen. But in all such cases the discoveries represented either an isolated fact or a single sequential relationship. (Well-known cases in point would be the discovery of the paralyzing power of curare, long used by the Amazonian natives; of the heart-stimulating power of digitalis, long recognized by the Plains Indians of the U.S.; and of the unrecognized discovery of Newton's laws of motion by the ancient Chinese who developed the principle underlying rocket propulsion.)

I.41 The great masses of verified knowledge underlying modern science—particularly its fundamental theories—have resulted from the cumulative efforts of well-trained and long-laboring men of superior intelligence who employed the methods treated in this volume. The inventions or discoveries sometimes made, even spectacularly, by laymen generally have very little meaning in and of themsleves. They must be fitted into an organized and coherent system of knowledge before they can attain significance or applicability. It is the laboriously constructed network of theories, principles, laws, facts and techniques, created by generations of scientists, which gives meaning to such discoveries or inventions occasionally made by laymen. Especially today, when the storehouse of scientific knowledge in the physical and biological fields is so voluminous and the theoretical structure so complex, it is highly improbable that an untrained person—no matter how innately intelligent—could make a significant contribution to modern sicence.

I.42 A final point of peripheral relevance and interest to the layman—in this discussion of the differences between science and pseudo-science—is the problem of choosing between conflicting claims made by reputable scientists. Scientists do not always agree; in fact, in some fields they even disagree strongly. The layman often wonders "Who's right?" This problem is inherent in the dynamic character of science, and is particularly acute in the less developed (particularly the social) sciences. New ideas, new theories, new findings, new interpretations of old findings— all these create honest and legitimate differences of opinion, even among objectively-oriented, well-trained investigators. Particularly in the applied fields of science (medicine or psychology, for example), where knowledge is far from complete, legitimate speculation ("hypotheses") or tentative theories may have to be substituted for reliable knowledge.

I.43 In such instances the layman has no way of solving his dilemma; in the language of metaphor, "You pays your money and you takes your choice." To the uninitiated who often thinks of science as a citadel of agreement ("Science says . . !"), it is well to remember that some of the greatest scientists of all time (e.g., Harvey, Semmelweis, Pasteur, Einstein) were at one time in their careers in sharp disagreement with the majority of their colleagues—let alone accepted by laymen. To the extent that a science is relatively undeveloped (sociology or linguistics, for example, in contrast to physics), then to that extent are its spokesmen less able to speak authoritatively in the language of reliable predictions; and therefore to that extent more imposters are apt to be found operating pseudo-scientifically in such a field. (A partial resolution of this layman's dilemma is the establishment of systems of public accreditation as has occurred in the field of medicine. Psychologists, for example, are moving to legitimatize through state licensing agencies the specification of who is and who is not qualified to practice psychotherapy.)

C. The Relation Between Science and Nonscience

I.44 Not all human behavior is guided by scientific knowledge. Most of an individual's acts are motivated by ideas unsupported by scientific evidence. In fact, the areas of greatest ideological involvement (politics, religion, morality, art, etc.,) are those wherein people function according to tenaciously held beliefs alone. Cogent arguments often are convincing in themselves, even without objective confirmation; and to many persons mere verisimilitude (i.e, the appearance of being true or real) is an adequate foundation for belief.

I.45 Everyday life consists largely of mental and physical habits based simply upon supposed certainty. "Common sense," for example, is regarded as an adequate guide for much daily behavior. Even the primitive person knows that water runs only down hill, that night follows day, and that all living organisms grow, reproduce and eventually die. The fact that much common knowledge is actually unconfirmed by scientific evidence is of little concern to the believer in the virtue of so-called common sense. If such a person believes that the stars determine human events (astrology), that probability in a card game is simply a matter of random behavior ("chance, luck"), that physical or racial features determine personality (physionogmic "psychology"), or that prayer causes rain—then such beliefs will often be accepted as personally and socially adequate guides to daily behavior.

I.46 Nonscientific knowledge, therefore, is essentially of three kinds: (a) that which is based mainly upon sheer habit alone (e.g., how to eat properly, how to ride a bicycle, how to drive a car, how to dance); (b) that which is accepted as factually true by one's reference group (e.g., certain "foods" are tasty, others

inedible; sex, race, nationality, class or occupational roles should
be sharply distinguished; there is a proper time and place for
such acts as eating, sleeping, etc.); and (c) that which is assumed
to be true by reason of particular ideological premises (e.g., religion is a "good thing"; democracy is the best form of government; *our* wars are always only defensive ones; monogamy is the
only morally defensible and naturally "logical" form of hetero-
sexual arrangement).

I.47 The often purported "conflict" between science and nonscience, therefore, assumes complex dimensions. The essential
element of credulity is *faith* ("I believe") or *certitude* ("I know").
To the believer, ghosts *do* exist, "a leopard *cannot* change its
spots" (i.e., personality *cannot* be altered), virtue *will* be re-
warded, and "luck" *will* change. And the faith or certitude of the
believer in science is essentially no different—i.e., is no more or
less credulous—than is that of the nonscientist. *Only the referent
is different.* The scientifically oriented person *believes* that the
bridge (built according to scientific "laws") *will* hold, that the
airplane *will* fly, that the drug *will* cure, and that the dead *cannot* arise. The referent in one case is faith in tradition, custom,
the consensus of one's peers, or in one's own powers of observation and reasoning. The referent in the other case is faith in a
method (viz., the scientific method) and its authorities. In cases
of dispute between the two explanatory systems, the deciding
factor for any given individual will be the certitude that either
system offers him.

I.48 For most modern individuals the problem of choosing
among different explanatory systems assumes confusing proportions. By the time one has reached a reasoning age, his knowledge
is an undifferentiated mass of habits, objectively verified facts,
and simple beliefs. Unable to distinguish clearly among the various types of knowledge he possesses, the layman often finds himself confused between opposing claims—of religion *versus* science
(e.g., Christ was born of a virgin *versus* parthenogenesis in animals
can result only in female offspring), of custom *versus* fact (e.g.,
wives should be younger than husbands *versus* women on the
average live longer than men), of science *versus* pseudo-science

(e.g., this illness can be cured only by drugs or surgery *versus* it can be cured by a magical panacea), or of ideological claims *versus* other ideological claims (e.g., private profit-motivated capitalism is the best form of economy *versus* socialism is the best form).

I.49 Aided and abetted by social training and indoctrination, this confusion is compounded by several factors. One such factor is expediency, or the need to "get things done." People cannot always wait for the scientifically verified answer to come in. Time may be "awasting," and the desire to act may be compelling. The common saying that "Something is better than nothing," often expresses such an impatience with the careful working out of a scientifically-planned analysis of a problem. ("Something is better than nothing," is not only specious logic but can also be fatal reasoning. If the "something" is a poisonous mushroom and the "nothing" is continued hunger pangs, it is doubtful that sure death is "better" than possible starvation.)

I.50 Another such factor is one's choice of faith-reference groups. A given person at any given time may believe in various —and possibly conflicting—authority figures (e.g., the pastor, the parent, the teacher, the public official, the editor of a newspaper, the leader of a gang); and often even such authority figures may all be woefully ignorant of the true facts of a given problem. Obviously the person who does not have the facts at his command is unable to judge objectively among conflicting claims to the truth. But when an argument is cogently presented from an approved reference group—as, for example, it so often is in nationalistic histories, in jingoistic newspapers, or in product advertising—then the layman is easily convinced that the knowledge so obtained is true because it is both credible and authoritative.

I.51 This state of confusion can easily be dispelled by the person trained in scientific method. Such a person learns, first, to distinguish and separate arguments of opinion from those of fact, and to realize that only the latter are resolvable according to scientific evidence. Secondly, he learns how to formulate a problem according to the demands of the scientific approach.

And thirdly, he learns how to assess conflicting authorities and how to sift and evaluate evidence. It is these three major abilities that distinguish the scientifically-oriented person from the lay-man. And it is essentially these three major processes that will concern us in the chapters to follow.

CHAPTER II

Scientific Reasoning

A. PHILOSOPHICAL ORIENTATION

1. The philosophical character of science
2. Science is a method, not a philosophy
3. Science and rationalism
4. Science and empiricism
5. The role of empirical evidence
6. Science and logical positivism
7. The status of neo-Positivism in science
8. Science and pragmatism
9. The value of pragmatism
10. Science and determinism

B. FACTS AND TRUTH

11. The notion of truth
12. Relation between facts and truth
13. What is a fact?

C. ASSUMPTIONS

14. Phenomena variously interpreted
15. The notion of self-evidence
16. Self-evidence examined
17. Knowledge and assumptions
18. Axioms, postulates, assumptions and presumptions
19. Postulate preferences
20. Postulates and scientific findings
21. The number of postulates in science

A. Philosophical Orientation

II.1 Science, as was noted earlier, is basically an objective, logical, systematic and verifiable method of analysis. But as has also been noted earlier, the scientific method is often placed in competition with other systems of analysis (e.g., spiritualism, dogmatism, intuition, revelation, authoritarianism). The question often arises, therefore, of the dependability of scientific knowledge, particularly when contrasted to its competitors or opponents. For this reason alone, the question is often raised of the philosophical character and the logical structure of scientific knowledge.

II.2 To reiterate, however, it should be borne in mind that science *is a method; it is not a philosophy* (of knowledge, of man, of nature, or of anything else, for that matter). Science as such, then—either as a method or as a body of verified knowledge— does not entail any particular epistemological position; that is, it is not committed to any particular theory or philosophy of knowledge. True, the scientist in action does exhibit certain mental preferences or consistencies in his method, and for this reason alone he has historically been labeled in various ways. Therefore, a brief examination of these labels and their philosophical implications might be useful in clarifying the scientist's position in relation to other students of phenomena who have somewhat different perspectives toward knowledge. It should always be borne in mind, however, that these particular philosophical labels (to be mentioned in the following paragraphs) denote what the scientist *does* when he reasons through a problem or examines data. They should not be construed as positions that he takes in regard to the inherent validity of one kind of knowledge *versus* another.

II.3 Historically and formally, "rationalism" refers to the philosophical conviction that human reason is both the chief instrument and the ultimate authority in man's search for the truth. Though not denying the value of sensory experience as a source of knowledge, rationalism contends that only the mind, operating logically, can determine the truth of any experience or idea. In the respect that the scientist adheres to established rules of logic, he might be labeled a rationalist; but the label would be misleading insofar as he does not trust "pure reason" alone as a guide to valid knowledge. The rational basis of scientific method is found in the system (i.e., rules or procedures) of logic employed in scientific reasoning; but the total method of scientific analysis requires much more than faith in reason alone.

II.4 "Empiricism," on the other hand, refers to the conviction that sensory experience should be regarded as the most reliable source of knowledge. Certainly science is in part and in certain areas an empirical method, just as it is a logical (i.e., rational) method. But the empirical aspect of science relates to the way data are perceived and not necessarily to a faith in the validity of sensory experiences alone. Reasoning about empirical impressions is just as important as are the sensations received.

II.5 Empirical evidence is basic to science, but it is meaningless in itself unless interpreted by particular notions about its attributes, its effects, etc. In fact a large part of the structure of scientific knowledge is composed of abstractions, not of empirical evidence—i.e., of ideas about phenomena and their interrelations ("theories" or "laws"). To say that science is empirical is really to say that the court of last resort (i.e., of establishing the reliability of any particular knowledge) is the empirical fact, the empirical demonstration, or the empirical prediction. But to contend that science is only, or basically, empirical is to invalidly limit its whole theoretical structure.

II.6 The former two philosophical "schools" (i.e., viewpoints) have a history extending back over three centuries; but a more modern "school" which has been linked to science is that of "logical positivism." Briefly stated, a logical positivist is one who believes that assertions have factual meaning only when they can be confirmed by empirical evidence. Actually a movement to establish a "Unity of Science" by some philosophers of science, logical positivism is mainly an attempt to unify various branches of science by clarifying ideas and developing precision of method through logical analysis. Although an outgrowth of empiricism, logical positivism stresses the development of objective communication (especially symbolic logic and mathematics).

II.7 To the extent that some scientists are moving constantly toward a common unity of method, basic principles and communication, then to that extent may one label some scientists as logical positivists. But among the relatively few "neo-Positivists" writing today, the original restrictive attitude toward the reliability of certain kinds of knowledge has been highly modified. Since the inherent limitations of Logical Positivism (i.e., as a formal philosophy) would exclude many modern and reputable scientific forms of analysis, this philosophy concerns us here only because of its influence upon all modern behavioral scientists. Incidentally, this "school" has only a tenuous connection with "Positivism", a philosophy of the nineteenth century which hoped to arrange all knowledge in a complete and cohesively interrelated organization for the rational solution of all human problems.

II.8 The fourth outstanding philosophical "school" attributed to science is Pragmatism—i.e., a belief that the ultimate test of the value of an idea is its usefulness in the solution of practical problems. Certainly the final arbiter of competing answers or solutions to any problem is the pragmatic test, "which one works," and to this extent scientists are often practical men seeking solutions to real problems. Later, in a discussion of validation in science, it will be noted that the "correct" (i.e., "valid" or "reliable") answer to a problem in science is often the pragmatic test of whether it achieves the results desired or predicted.

II.9 But as a philosophical position, pragmatism is of little value in modern science. Much knowledge in science is purely theoretical, and hence not of pragmatic value *per se*. Yet it plays a very vital role in the theoretical structure of any valid science. In fact, so-called "pure" science—i.e., the abstract theories which underlie all science—are highly unpragmatic (i.e., are not testable directly). This point will be developed later in Chapter V when discussing the role of theory in science.

II.10 The fifth and final philosophical linkage to science is "determinism"—i.e., the contention that nothing takes place in nature without natural causes. Actually a "postulate" rather than a "credo" (see the discussion of postulates later in Section D), the "deterministic assumption" is employed in science largely in the analysis of causation—a topic to be treated at length in Chapter VIII. Once allied to the philosophy of "materialism" —i.e., the doctrine that all knowledge can be derived from a study of matter—today science is materialistic, mechanistic or deterministic (these concepts are somewhat related but certainly not synonymous) *only* insofar as it rests upon a foundation of objectively demonstrable facts as evidenced by instruments of a physical sort (meters, gauges, dials, etc.). Determinism, further-more, should not be connoted as "fatalism," or the natural in-evitability of events. Science seeks to understand regularities in phenomena, but such regularities are not imputed to any in-evitable causative agent. A "postulate of regularity in nature" is assumed by the scientist as a working principle in order to achieve reliable knowledge; but such a working principle is not assumed to be a so-called "law of nature." This latter term has no significant meaning in modern scientific explanations of causation.

B. Facts and Truth

II.11 It is commonly assumed that science deals only with "facts," and that its basic function is to search for and reveal the "truth." Such a view, though basically correct, vastly over-simplifies the scientist's problem. The notion of truth alone— excluding such allied notions as "valid," "verified," "proved" or "demonstrated"—has plagued epistemologists and other philos-ophers for centuries. The difficulty of defining the term "truth" stems from the assumption that something either is basically, inherently, or necessarily, so, or that it is not so. Yet the history of human experience has demonstrated quite clearly that what was once held as unquestionably true (e.g., that the sun revolves around the earth) was later just as unquestionably deemed to be false. Furthermore, at any given time different groups will de-fine the same (or at least apparently the same) phenomenon quite differently. To some persons, for example, the Virgin Birth of Christ is an indubitable fact; to others it is an erroneous and hence superstitious belief. To some people it is unquestionably true that criminals are "born bad," or that Orientals are "naturally" crafty; while to others such notions have no substantial evidence to support them. This difficulty (of defining "truth," due to the notion of *inherent* truth or falsity) is—as will be explained below—avoided in science.

II.12 Science is based upon *facts*—i.e., upon truth assertions ("propositions") supported by objective evidence. Such assertions, however, do not imply that a fact is *necessarily, inevitably,* or *inherently* so. It *is* a fact merely because substantial objective evidence seems to support it. Truth in science is never final or absolute; it is relative to the amount and kind of evidence which substantiates it. (See Postulate (*8*) later in this chapter.) The reason why all factual knowledge in science is relative, rather than final or absolute, is a consequence of its experiential character. Facts derived from experience lead to *probable* truths, never to *certain* truths, for *experience is infinite*—i.e., future experience can always require a new interpretation of a phenomenon. But since frames of reference may differ (as noted in the previous chapter) even among scientists, it is not surprising to find occasional disputes regarding the validity of an asserted fact. According to one "school" of psychology, for example, it is a fact that dreams are evidence of subconscious desires. But according to the critics of such a viewpoint, specific dreams (i.e., the evidence) do not substantiate the "fact" that subconscious motives are responsible for dream content.

II.13 "What is a fact?" therefore, depends upon the acceptability of the evidence offered. Some factual statements are supported by unquestionable objective and empirical evidence (e.g., the earth moves around the sun in highly regular cycles), while others are supported by much less convincing evidence (e.g., smog is a cancer-causing agent). Some factual statements are dubious in terms of scientific evidence (e.g., men are more logical than women); while others are unquestionably false (e.g., wearing amulets insures preferential treatment from natural forces).

Thus it can be seen why the original question of "What is the truth?" sometimes receives the answer: "Well, it depends . . ," (upon the amount and kind of evidence deemed adequate by the person questioned). Further exposition of this problem of fact and truth will be deferred until a later section in this chapter.

C. *Assumptions*

II.14 Though a scientific fact may be defined as a truth statement (or "proposition") supported by convincing objective evidence, the reader already may have wondered why different persons can observe the same phenomenon and yet derive different interpretations of it. (Imagine, for example, an aborigine and a modern astronomer both interpreting an eclipse of the sun.) The interpretations of either single or related phenomena (i.e., "descriptions" or "explanations") are usually based upon some prior knowledge presumably related to the particular phenomenon in question. Therefore, since some facts are required in order to prove other facts, all knowledge systems are eventually driven to prove those basic facts which are employed to prove other facts. But the obvious question immediately arises: How are the basic facts themselves to be proved? The answer most often given is "by self-evidence." That is, the basic or fundamental facts which underlie all knowledge systems are presumed to be a kind which need no further proof because they are "obvious" to "any intelligent" person.

II.15 Thus to ask "How do you know that you have a toothache?" might evoke the answer "Because it hurts!" To the next question "But how do you know it hurts?" the most common answer would be something on the order of "Because I just do, that's why!" It is interesting to note in this illustration that the victim of the toothache finds it very difficult to prove objectively to anyone else—and particularly to one who has never had a toothache—what seems to him so obviously to be a fact. But there is no question in the mind of anyone who has had a toothache that the fact of the pain is obviously and self-evidently "true."

II.16 Self-evidence, however, is a dubious and often unreliable basis for the establishment of valid knowledge. What is obvious or self-evident to one person may appear highly dubitable or even erroneous to another. At one time many "intelligent" persons believed as self-evident such basic "truths" as the notion that the sun revolved around the earth, that spontaneous generation occurred, that some people could arise from the dead, that rain was the tears of the gods, that ghosts existed, that man could reason but animals couldn't, or that witches were demons in human form. It would be both provincial and dogmatic to assume that "the others" were obviously wrong because now "we *know*" what the truth in such matters *really* is.

II.17 Underlying all knowledge are certain basic *assumptions* vital to the reasoning process. Such basic assumptions are termed "axioms" or "postulates"—i.e., beliefs accepted not only as true ("axiomatic") but also as necessary prerequisites to any argument or discourse. Put another way, it may be said that some framework of reasoning, some basic agreements, are necessary to any logical or epistemological system in order for persons to interact at all. In this sense, postulates can be viewed as the "rules of the game" (of interpreting evidence and thereby of deriving knowledge). The postulates of science, therefore, are the basic assumptions about experience which scientists employ to interpret the evidence necessary to produce verified facts.

II.18 Strictly speaking, there are substantial differences between axioms and postulates. An *axiom*—as the term is to be used here—is an assertion about reality (i.e., a "proposition") accepted without proof because it is assumed to be more or less self-evident (e.g., if A is equal to B, and B is equal to C, then "obviously" A is equal to C). A *postulate*, however, in this context, is a proposition which is offered, and presumably accepted, as true without further proof simply because some sort of agreement is necessary in order to communicate at all. While on this topic of definitions, it might be useful to clarify two more terms which are often employed in this context of basic reasoning devices. A *presumption,* for example, is a proposition supported by probable, though not conclusive, evidence or proof; and which is accepted as true because of its reasonableness or high probability. An *assumption,* on the other hand, is a proposition that is neither self-evident nor highly probable. Nevertheless, it is offered so that it may serve as a premise in a particular discourse. To summarize the preceding paragraphs, therefore, it may be said that science employs all four of these types of propositions at different levels of reasoning in order to develop various stages of communication. At the moment we are concerned with only one of those stages: the role of postulates.

II.19   The scientist's preference for a particular set of postulates over others is based primarily upon their functional superiority in terms of his basic purposes: viz., to understand phenomena in order to predict and possibly even to control them. In the first place, the postulates of science demonstrate a logical consistency not offered by other belief systems; in fact, many of them appear (even to the layman) "common-sensical." In the second place, they permit an orderly system of inferences about the kind of phenomena with which the scientist is concerned. In the third place—and this is their most significant feature— they seem so far to have provided more dependable answers to the kinds of questions the scientist asks. This latter feature alone, (and here we see in operation the pragmatic test of utility) decrees that these postulates are preferable to others simply because they seem so far to have produced the kinds of results the scientist is attempting to achieve.

II.20   Two more features of scientific postulates should be borne in mind. First, they should not be confused with the *findings* of science. The postulates of science might be changed in time if new knowledge should demand new frames of reference, for new knowledge often changes the status of previous scientific findings. But the one is not necessarily dependent upon the other. The postulates of science exist as *functional* devices useful for the job to be done, whereas scientific findings are confirmed by objective experiential evidence. Second, the postulates serve to support the basic attributes of scientific method previously discussed by *providing the rationale* for those attributes. Thus the attributes and the postulates form an interwoven system within which the postulates are basic intellectual *implements* necessary to the pursuit of the kind of inquiry that concerns the scientist. It is this entire system which underlies scientific method, and which to the scientist seems to demonstrate a vast superiority over any other present system of explaining man and the universe which surrounds him.

II.21 A final note about the following listing of postulates should be made. Examination of a wide variety of books on science would reveal that no agreement yet exists regarding either the number or the designation of such postulates. Many writers refer to two or three "basic" postulates, others to five or six. Some writers infer such postulates but never clearly designate them; and many writers fail to mention any specific postulates at all—probably because it is assumed that "everyone" knows them. The specification of ten particular postulates, therefore, should not be construed as representative; for representative or typical treatments of scientific method do not as yet exist. This listing is, rather, an attempt to concretize and aggregate what seem to be generally accepted among competent authorities as the essential assumptions underlying scientific method. As such, they represent this writer's contribution to the possible standardization of such listings for purposes of future discussions.

D. Postulates

II.22 *(1) All behavior is naturally determined;* i.e., all events have a natural antecedent ("cause").

This postulate epitomizes the great historical break of modern science away from fundamentalist religion, on the one hand, and from spiritualism and magic on the other. It implies, in effect, that explanations of events shall be sought in natural causes or antecedents. (The term "natural" will be defined and examined shortly.) It eschews supernatural definitions of phenomena, and rejects the notion that forces, agents or agencies other than those found in nature operate to influence the cosmos, the earth and its flora and fauna. When a supposedly supernatural or extra-natural explanation is offered for a perplexing phenomenon, the scientist assumes that the answer will be found in natural forces or events. And until such time that he can explain the event in natural terms, he rejects the belief that some other order of explanation is necessary. If the history of science proves anything, it proves that the scientist has not yet had his confidence in this belief shaken by developments to date.

II.23   This postulate is not "deterministic" in the sense of implying inevitability. Its main function is to direct the search for "causes" away from whimsical or omnipotent agents (i.e., gods, spirits, etc.), and toward the regularities that apparently underlie many phenomena. Like another postulate (treated later) about the regularity of natural phenomena, this one simply serves the function of assuming that behavior is not either wholly random or unexplainable. To the believer in a Divine plan or purpose, this postulate can easily be subsumed as a consequence of such a plan or purpose. Hence it is quite reasonable to find that many theologically oriented persons can effectively employ the scientific approach based upon this postulate. Such persons simply assume or believe that God—however He may be defined —is both the Creator and Cause of all natural phenomena.

II.24   The concept "nature" or "natural" is probably one of the most abused terms both in serious intellectual discussions and in common speech. It might be useful at this point, therefore, to define it when it is accurately employed in science. Essentially, it denotes all those objective and empirically demonstrable phenomena which exist independently of man's intervention (but including man himself, of course, as a biological entity). To avoid possible confusion about the implications of this postulate of causation (sometimes referred to as a "heuristic" or explanatory device by philosophers of science), it might be helpful to define the concept "natural" in terms of what it does *not* mean when properly employed. It does not refer (a) to habit patterns ("Naturally he stepped on the brakes"); (b) to indiscriminate regularities ("Naturally he took the easiest way out"); (c) to an omnipotent force ("Mother nature descended in all her fury with the storm"); (d) to established preferences ("Naturally he took the candy instead of the spinach"); (e) to presumably normal behavior ("Naturally like any man he enjoys poker more than sewing"); (f) to simply unique behavior ("The child's painting was natural and unspoiled by formal training"); (g) to personality constitution ("It is his nature to be proud and vain"); or (h) to the intrinsic qualities or peculiar attributes of (1) an object (e.g., "The natural form of a wheel is round"); (2) an animal (e.g., "Dogs naturally like to chase cats"); (3) a person (e.g., "John naturally likes to swim"); or (4) a phenomenon (e.g., "Competition naturally is more effective than cooperation"). In short, the concept of naturalness, while including man, refers to objects, conditions or events (viz., phenomena) which exist or operate independently of man's manipulation of them due to forces existing beyond his creation of them.

II.25 (2) *Man is part of the natural world;* individual and social phenomena can be understood by objective methods of study.

This postulate is basic to the development of an objective and reliable social science. Essentially it avers that man is just as much a part of nature as is any other animal; and although he possesses distinctive attributes and therefore exhibits distinctive problems of analysis, he can nevertheless be studied by the methods common to all the reliable sciences. The tenacious distinction made by laymen between so-called inherently subjective and inherently objective phenomena implies that man cannot be studied objectively because he possesses a "soul" or "will" or "mind," or even because of his uniqueness ("individuality").

II.26 This popular distinction is avoided in the more advanced social sciences. However different man may be from other animals, the modern social scientist assumes that man exhibits enough recurrent and observable behavior to be able eventually to be better understood than he has been in the past or by the use of other methods of analysis (e.g., astrology, biological determinism, revelation). Although man undoubtedly is unique both as a biological species and as a member of a particular cultural group, nevertheless the commonalities inherent in group life permit him to be studied scientifically. Many specific problems involving this postulate will be treated later when viewing the scientific method employed in the social sciences.

II.27   (3) *Nature is orderly and regular;* events do not occur haphazardly.

Inherent in the scientific analysis of natural phenomena is the belief that the universe operates according to certain patterns of regularity (viz., so-called "natural laws"). It should be pointed out, however, that this regularity is only *apparent* from objective empirical experience, it is *not* assumed to be *necessarily* inherent or innate in natural phenomena. That is, there is no factually inherent or logically necessary reason why, for example, the sun should rise each morning, or why spring should follow winter; but they do, and repeatedly so; and this regularity seems to underlie all known natural phenomena and their interrelations. In practice, this belief takes the form of explanations expressed in terms of probabilities inferred from the particular to the general, or from past experiences to the present and hence to the future. That is, it explains repeated occurrences by simply stating that, according to previous observations, the probability that the sun will rise tomorrow morning is, let us say, several billion to the one that it will not. But it also concedes that absolute invariabilty (i.e., inevitability) might never occur in fact—e.g., there is no absolute assurance that the sun will continue to rise every morning. Conversely, this postulate rejects the notion of purely random or unrelated occurrences, and directs the attention of the scientist toward seeking the qualitative and quantitative relationships which apparently exist between and among natural phenomena.

II.28   This postulate of regularity in nature has an important implication in the explanation of phenomena. It implies, for example, that nothing occurs by pure chance (e.g., "luck," "the breaks," "fate," "destiny"). In fact, when the term *chance* is used in science (as it is in probability statistics), it refers simply to unknown and therefore unpredictable relationships, never to uncaused or inherently random events. According to this postulate every occurrence has a necessary antecedent; and although many phenomena may appear unique (e.g., no two snowflakes are ever identical in form, and no two tornadoes ever seem to follow an identical pattern), nevertheless underlying such unique or unexplained events are certain patterns of forces which, when understood, will permit better prediction than would be possible by sheer guesswork alone. Though allied to the first postulate (viz., of natural causation), this one goes beyond it to contend that nature is not whimsical or capricious; but if "she" appears to be so, it is only because we do not yet understand (i.e., have not yet found) the regularity which presumably is there to be found.

II.29   (4) *Nature is uniform;* not infinitely complex.

This postulate expresses an observation of apparent fact, and permits science to classify data into *taxonomic* (i.e., functionally arranged classificatory) systems. Such systems are logically necessary if the large masses of facts are to be handled in any kind of manageable order. But it should be pointed out that such systems are artificial mental creations—i.e., they do not exist naturally apart from the qualities given them by man. (This is simply another way of saying that man creates categories to serve his purposes; after all, animals do not have to be classified as "mammalia" simply because they suckle their young.) Although the dividing lines between different orders or classes of phenomena may at times appear quite indistinct, nevertheless the general, average, or essential differences between such classes are real enough to permit functional generalizations to be made about them. In practice this postulate also permits the scientist to develop theoretical analyses of interrelationships between classes of phenomena, and from these to proceed to the larger analysis of the cosmos as a whole.

II.30   These last two postulates (viz., order and regularity, and uniformity) are usually combined by most writers under the general nomenclature of the *uniformity of nature.* They are separated here simply for convenience of treatment, but should be viewed as a composite whole. Their combined implications form the foundation of scientific logic when applied to natural phenomena. Furthermore, they not only permit generalizations about and classifications of phenomena, but they underlie all probability theory; and therefore are indispensable to all sampling procedures. Finally, these combined postulates suggest the increasing possibility of achieving a more highly integrated, general theory of explanation which—in conformance with the principle of parsimony—is a basic goal of all scientific endeavor.

II.31   *(5) Nature is permanent;* although all things apparently change in time, albeit at varying rates, many phenomena change slowly enough to permit the accumulation of a reliable body of knowledge.

This postulate is a logical necessity in any prolonged effort to accumulate reliable knowledge. What it implies in effect is the belief, for example, that a lump of coal studied today—although perhaps indeterminately changed by tomorrow—will, nevertheless, be similar enough to permit valid generalizations made about it to hold true for a given period of time. If this postulate could not be believed, it would be impossible to know anything except for the brief moment that the perception was taking place. In practice it underlies the cumulative attribute referred to previously.

II.32   The concept of the permanency of nature is, like most postulate concepts, relative rather than absolute. What it refers to essentially is a general feature or common tendency, not an inviolable, never-changing fact. In the realm of personal and social behavior, for example, it is obviously true that changes in phenomena often occur quite rapidly; but it is just as undeniably true that even in social behavior many general and relatively consistent patterns of phenomena can be demonstrated. The major utility of this postulate lies in its promise of permitting the accumulation of verified and predictable knowledge; and its major justification lies in an observation of apparent fact deduced from experience with one's environment.

II.33   *(6) All objective phenomena are eventually knowable;* given enough time and effort, no objective problem is unsolvable.

This postulate stems from two related convictions: (a) that human intelligence is capable of unlocking the secrets of the universe—i.e., a faith in the power of human reasoning; and (b) that man's search into the mysteries of objective phenomena has so far been so fruitful that apparently no doors to knowledge are immutably locked to the continued efforts of scientific search. As much an article of faith as an operational principle, this postulate rejects the notion that there are any intellectual limits placed by nature upon man's search for knowledge.

II.34 No matter how unpopular or apparently unfruitful his curiosity may be, the scientist regards all objective phenomena as worthy of study. In reality an intellectual conviction as well as a functional necessity, this postulate expresses the belief that —since no one can know beforehand where inquiry will lead— the scientist must be prepared to travel all possible avenues in the search for knowledge. It is quite true, of course, that a scientist may limit his endeavors in terms of moral or ethical considerations (e.g., conducting experiments that knowingly would require murdering innocent humans), or in terms of particular preferences amongst various interests. He does not, however, as a scientist arbitrarily limit his search for knowledge in terms of popular interests or notions of propriety. In short, he does not assume that just because a particular topic of interest is unpopular (e.g., sexual aberrations, deformity, loathsome diseases, graft and corruption), that therefore it is unworthy of scientific study. Incidentally, the history of science suggests that it has been partly on this account that scientists have often been denounced and even persecuted, not only in the distant past but occasionally even today.

II.35 (7) *Nothing is self-evident;* truth must be demonstrated objectively.

This postulate avers that reliance should never be placed upon so-called common sense, tradition, folk authority, or any of a number of customary interpretations of phenomena. The scientist well knows from historical example as well as from his own experiences that apparent veracity is often quite different from objective, empirical verification. He also learns early in his career

that, statistically speaking, there are many more possible wrong
answers than right ones; and that (to paraphrase mythology), the
devil in the form of error often appears seductively tempting. It
is for this reason that skepticism is an occupational attribute of
the professional scientist, and even the simplest of notions often
is resubmitted to objective verification.

II.36   There is another feature of this postulate that should
be mentioned at this point. Postulate systems may serve several
purposes: (a) They may provide an ideological or moral framework for belief systems. (b) They may serve as orientation points
in the development of theory. (c) They may fill in gaps in existing knowledge by providing presumptions rather than facts. In
science this particular postulate admits that gaps exist in the field
of knowledge desired about the universe; but it avers that beliefs should constantly give way to objectively demonstrable
facts. In this third sense alone, postulates serve the scientist as
reference points for viewing the phenomena he is studying, even
though particular postulates might eventually be replaced by
verified facts or by different postulates. The first two functions
herein mentioned, however, would still remain to serve in the
development of any body of knowledge.

II.37   (8) *Truth is relative* (to the existing state of knowledge);
absolute or final truth may never be achieved.

Contrary to popular thinking and to the orientation of early
philosophers, the scientist soon learns that knowledge is dynamic.
This postulate does not imply that no stable knowledge can
ever be acquired; but it does recognize the fact that as knowledge
grows in quality (i.e., becomes more highly verified) as well as
in quantity, reinterpretations of (and therefore conclusions
about) phenomena become imperative. This open-mindedness so
characteristic of scientists is the outstanding attribute which has
permitted science to grow as spectacularly as it has, and which,
at the same time, has encouraged the constant re-evaluation of
old as well as of new ideas. Again, contrary to fixed systems of
thought, proof in science is always relative: to the time, the data,
the methods, the instruments employed, the frame of reference,
and therefore to the interpretation. In this respect, truth in sci-

ence is simply an expression of the best professional judgments demonstrable at any given time.

II.38 *(9) All perceptions are achieved through the senses;* all knowledge is derived from sensory impressions.

The student of philosophy and psychology would be quite familiar with the historical controversy centering around the so-called "sensationalist" school of the eighteenth and nineteenth centuries, which contended that the only reliable experiences were those of the senses. This postulate—although related to that controversy—extends outward from it; and in a sense goes beyond it to aver that the only reliable knowledge is that which is both objectively and empirically verifiable. This postulate rejects, for example, the notion of extra-sensory perception as well as of divination, thought transference, and other types of intellectual stimuli which presumably impinge upon man's awareness. As will be noted later, this postulate should not be construed in the narrow definiton of the five common senses (i.e., touch, taste, smell, hearing, seeing); but it should be interpreted to mean that the materials of reasoning (viz., ideas, concepts, constructs, images, etc.) are in part fashioned from the impressions received through the senses.

II.39 This so-called "postulate of empiricism" is largely a consequence of a major shift in Western science which occurred roughly around the beginning of the seventeenth century. Up to that time the tradition of Greek thought, especially the influence of Aristotle, had placed almost complete credence in "pure reason" as the major human ability for the analysis of phenomena. The monumental influence of Galileo (at the turn of the seventeenth century), which is second only to that of Newton, who followed him, was partly due to his insistence upon empirical demonstrability of theoretical predictions. It was Galileo's impressive union of theory with experimental verification that determined the course of modern science. Though modern physical science, as a synthetic system, is held together by a core of theoretical notions (viz., theories, laws, principles, etc., about the universe), it is the empirical fact that lifts it beyond the range of speculation.

II.40   The postulate of empiricism is often questioned in regard to so-called extra-sensory perception—the so-called "sixth
sense." If true extra-sensory perception should ever be objectively
and empirically demonstrated to exist, two alternatives are possible: (a) the postulate could be abandoned as unreliable; or (b)
the definition of sensory impressions would have to be broadened to include others not now recognized. A somewhat comparable situation occurred several generations ago when, according
to informed ornithologists, the definition of a swan began "A
swan is a white bird which . . ." Then black swans were discovered in Australia. This discovery did not invalidate the original definition; a simple change broadened the original concept of
a swan to "A swan is a bird, usually white, but which may be
black when native to Australia, which . . ." (The role of definitions in science, as illustrated partly by this example, is so significant that a major treatment of it is deferred until a later
chapter.) This postulate, then, expresses a fundamental conviction, following the tradition of Galileo, that empirical demonstration is the ultimate test of the validity of all theoretical speculations about objective phenomena and the resultant predictions.

II.41   *(10) Man can trust his perceptions, memory and reasoning* as reliable agencies for acquiring facts.

Contrary to some philosophers—who denied that man is able
to trust his perceptions or his reasoning—the scientist has faith in
his sensory abilities and in the intellectual interpretations he
makes of their impressions upon his mind. Furthermore, he believes that although human reasoning is fallible, nevertheless it is
the only means he has to interpret the world about him. Today
few persons would question this postulate seriously; but nevertheless it underlies the whole rational and empirical basis of
scientific knowledge.

II.42   This postulate does not imply that any and all perceptions, memories or reasons are reliable. Any student of psychology well knows how unreliable sensory impressions, recall,
or reasoning may be. But what this postulate does aver is that
the final resolution of any dispute about phenomena should be
based upon accepted rules of reasoning and upon sensorily perceived data—not upon mere mental notions or ideas. Further-

more it does not imply that any particular perception must be trusted; but simply that ultimate trust in one's analysis of phenomena must be based upon empirical evidence as interpreted according to accepted rules of logical reasoning.

E. The Role of Logic

II.43   Though verified facts are the building blocks of science, they must be assembled and arranged into useful and interrelated structures. Contrary to popular belief, facts do not "speak for themselves," nor is it necessarily true that "facts don't lie." (Actually, such statements have no substantial meaning in themselves.) The most essential tool of science, along with the verified fact, is the system of valid reasoning ("logic") about facts that permits reliable conclusions to be drawn from them. Such conclusions are variously termed "theories," "principles" or "laws" —i.e., propositional statements about the interrelation of facts explaining a given phenomenon.

II.44   At the core of logical reasoning about facts is a system of rules and prescriptions which have been established over the course of twenty-five centuries of Western thought. The correct employment of such rules is basic to all scientific effort. The rules of "deductive" and "inductive" inference, of the correct use of definitions, of sampling procedures, etc., are essential parts of the intellectual tool kit of any scientist. In this last section, we shall briefly examine only one of the major logical problems involved in scientific reasoning, to wit: the distinction between truth and validity. This choice is made in order to illustrate the very significant interrelation between facts (i.e., truth statements) and the logical arrangement of those facts (i.e., valid reasoning) which comprises the theoretical core of the scientific structure.

II.45   Most errors in reasoning occur from the common tendency to confuse truth with validity. A fact is either certainly or probably true when substantial objective evidence exists to support it. But an argument (i.e., a claim) is valid only when the conclusion necessarily follows from the assumptions ("premises") made originally. A person can easily draw a wrong conclusion from verified facts if he reasons incorrectly; but he can just as

easily draw a wrong conclusion by reasoning correctly if he employs incorrect "truths" as premises. The key to this problem of
logical inference, therefore, is the *mode* of the valid argument.

II.46 There is only one correct ("sound") form of argument
accepted in logic—viz., one wherein the premises or assumptions
are factually true and the inferences drawn from them are valid;
but there are three forms of incorrect ("unsound") argument.
Perhaps this ratio may help to explain why reliable (i.e., highly
dependable, predictable to a high degree) knowledge grows so
slowly in so many fields. And it may also suggest why so many
persons often draw incorrect conclusions even from verified facts.
The four forms (viz., one sound, three unsound) are worth illus-
trating at this point.

II.47 Let us consider this argument from botany: A poppy
is a plant, and all plants need moisture to live; therefore a poppy
needs moisture to live. The assumptions or premises (i.e., about
a poppy being a plant, and about plants needing moisture to
live) in this case are both factually true as established by experimental evidence. The conclusion, therefore, which links one
plant (viz., the poppy) with all plants, "necessarily" follows, and
therefore is "valid," thus making this a sound argument. The
reason the conclusion *necessarily* follows is inherent in the contentions of the premises. The first one identifies a member of a
class of objects (viz., poppies), while the second one states a condition pertaining to all members of that class (viz., that they need
moisture to live). Any reasonable person, therefore—if he accepts the normal meanings of the language employed (e.g., of
"needing moisture to live")—is bound to arrive at the deduced
conclusion.

II.48 Now let us consider a second form (and the first of three
unsound forms) of the same argument. A fresh poppy is combustible, and all combustible things can be burned; therefore a
fresh poppy can be burned. This argument is unsound because
the first assumption (or "major" premise) is false in fact—fresh
poppies are not combustible. But if they were in fact combustible,
then this would have been a sound argument; for the second assumption (or "minor" premise)—viz., that all combustible things
can be burned—is true by definition alone. Therefore, the con-

clusion from that standpoint is valid though untrue. This example should clearly illustrate how easy it is to draw reasonable (i.e., valid) conclusions that are untrue simply because at least one of the facts was not true in the first place. This example could be compounded by the case where both assumptions are factually wrong, but if accepted, would lead to a valid (though unsound) conclusion. Consider, for example, this argument: Poppies contain vitamin x and vitamin x prevents baldness; therefore (eating, or synthesizing and then ingesting) poppies prevent baldness. The argument is valid because the conclusion necessarily follows from the premises; but it is unsound and therefore unacceptable because in this case both premises happen to be false.

II.49 A third (and the second unsound) form of argument might be illustrated thusly: Poppies need oxygen to live, and humans need oxygen to live; therefore poppies are human. In this case, both assumptions happen to be true in fact, but the conclusion (viz., poppies are human) is invalid. The reason why it is invalid is worth stressing. The argument does not include in its assumptions the contention that things which have something in common (in this case, needing oxygen to live) necessarily are alike in other respects. After all, fish swim and men swim, but this common ability does not necessarily make them alike in other respects. Or, as one "F" student once wrote in a freshman logic course examination, "Men wear pants, and lamb chops wear pants; therefore men are lamb chops."

II.50 The fourth (and third unsound) form of argument might be stated somewhat like this: Poppies are succulents, and succulents have thorns; therefore poppies have thorns. This argument is unsound because it is both invalid in form and false in fact. Both premises are known to be false in fact to begin with; and the conclusion—i.e., what might be true of some members of a class (viz., succulents having thorns) therefore is true for all members of that class—does not necessarily follow. Though such an unsophisticated argument never appears in any reputable science, it is interesting to note in all these examples how easy it would be to change the structure of the arguments from sound to unsound, or *vice versa,* simply by altering them slightly. Most often this is done by assuming something to be true which actu-

ally is not so in fact; or by deducing various forms of the old
high school geometry axiom that "things equal to the same thing
are equal to each other" (which they aren't necessarily).

II.51   The above illustrations about poppies could be ex-
panded to great length, not only in popular reasoning but also
from the history of science itself. Without belaboring the point,
a few classical examples may help to illustrate the fundamental
role that verified facts and sound reasoning play in the structure
of science. These examples are chosen at random: In pre-Galilean
physics, it was contended that heavy objects (necessarily) fall
faster than light ones. This contention is false in fact and invalid
in form. Heavy objects fall faster than light ones only when
they display a lower unit-volume ratio of resistance to the medium
(i.e., liquid or gas) in which they are falling. A one-pound ball of
lead will fall much faster than will a ten-pound bag of noncom-
pressed feathers; its unit-volume of resistance is smaller. Not
only is one of the assumptions in this argument false in fact (i.e.,
that absolute weight, rather than density, determines the rate of
fall), but the conclusion is invalid because it does not necessarily
follow that just because things are heavy, they necessarily fall fast
(i.e., faster than light things). The conclusion also can be shown
to be false in fact (as Galileo purportedly demonstrated from the
Leaning Tower of Pisa, although this report is also false in fact),
in a simple laboratory experiment.

II.52   A different type of example might be drawn from social
phenomena, to wit: Negroes have a proportionately higher crime
rate than do Whites. This common contention (made even by
chiefs of police, prison officials and legislators) illustrates several
facets of faulty logic: (a) It implicitly infers that *all* Negroes have
a higher potential for committing crimes; and this is not a fact—
only some classes of urban Negroes in the U.S. exhibit a higher
rate of some types of popularly-noticed crimes. (b) It also infers
that Negroes—even if they should in fact commit proportionately more crimes—do so *because* they are Negroes. This implication is also false in fact, as any social psychologist well
knows. Without elaborating this example, it is interesting to note
that here is a case of a partially-true assertion which is valid for

the wrong reason (i.e., wrong in the sense of popular attribution or causative reasoning). In the cases where Negroes have a proportionately higher crime rate than do Whites, they apparently do so because (1) crime rates are proportionately higher in urban than in rural areas; and the Negro-criminal allegation is largely an outgrowth of Negro immigration into urban areas; (2) popularly-noted crimes are generally of a lower criminal class variety (e.g., rape, robbery, burglary and assault); and Negroes predominate in the lower classes due to the racial discrimination exhibited against them in jobs, housing, etc; and because (3) crime rates as indicated by the population of penal establishments (where Negroes are proportionately high) simply reflect disproportionate social (i.e., police, the courts, the press) sensitivity to lower-class crimes. There is no substantial evidence, for example, that Negroes commit proportionately as many (let alone more) upper-class crimes (e.g., embezzlement, big-time gambling, forgery, espionage,) than do Whites; in fact, growing evidence strongly suggests that they commit proportionately fewer of such kinds of crimes. If, however, it is clearly established as a matter of fact that one's criminal-behavior potential is in no way a direct and necessary consequence of his racial inheritance, then it may be validly deduced that—if and when Negroes achieve complete social, economic, political and legal equality with Whites—they probably will commit the same proportion of all types of crimes as will Whites.

II.53   The foregoing examples should suggest the fundamental role that sound reasoning plays in the scientific approach. Needless to say, no person can become a competent scientist—although he may become an effective practitioner—without a thorough grounding in logical reasoning. It is not within the scope of this volume, however, to attempt such a task; our concern is with the role that facts, and reasoning about facts, plays in science. In the chapters to follow, both the philosophical basis of science and logical reasoning will be illustrated in visualizing the total structure of scientific method.

CHAPTER III

Problem Formulation

A. PROCEDURAL STAGES

B. FEASIBILITY OF SOLUTION

A. Procedural Stages

III.1   All scientific inquiry stems from a problem—e.g., *Who* will vote for which candidate? *What* types of plants respond to this treatment? *Where* is corrosion most likely to occur? *When* will the train arrive? *How* is social class related to age at first marriage? Indirectly, in all these questions, a *Why?* may also be implied; but unlike the basic queries mentioned, a "why" question occupies a special place in science. This special place is a consequence of the implications underlying the kinds of questions asked above.

III.2   The first five basic questions simply asked for a determination of discrete connective relationships—i.e., which phenomena are connected, when are they connected, etc., to other phenomena. But the last question ("Why?") asks for a different and more complex kind of relationship between phenomena—i.e., a relationship expressing a *causal* connection. This type of relationship is so significant in all inquiry, scientific or not, that it is dealt with at length in Chapter VIII. For present purposes, however, let us say that a problem about causal connection in science is eventually answered in terms of "necessary" or "sufficient" conditions which relate a given "cause" to a given "effect." (To say it in other words, the answer states conditions which relate an antecedent event to a consequent event.) But such "necessary" or "sufficient" conditions which explain the cause of a given effect—and thereby answer the question of "Why?"—are essentially statements of Who, Where, When, or How conditions. The initial stage in the scientific approach, therefore, is a specific inquiry seeking the connective relationships between two or more classes of phenomena.

III.3   Once the initial inquiry is posed, the systematic structure of the scientific approach becomes immediately evident. For, unlike random inquiries, an inquiry in science is structured in a rather precise and logically arranged form. This form has been developed over the centuries from a wealth of experience in answering scientific questions. The procedural and structural aspects of the scientific approach are, in a real sense, synonymous.

This procedure, as will be shown below, determines rather precisely the various sequential stages through which one progresses in order to answer his query in a manner satisfactory to the scientific community.

III.4 From the initial query to its final answer, a scientific approach proceeds through *eight* major stages of operations. The *first* stage is the formulation of the problem by a statement of an empirically testable proposition ("hypothesis") expressed in objective or operational terms. It is at this point, it should be noted, where the logical theoretical structure of science has so large a role to play—namely, in suggesting meaningful hypotheses.

III.5 The *second* stage is a study of the pertinently related literature for any assistance it might offer in terms of data or methods of procedure.

III.6 The *third* stage is the construction of a research design by which the problem is to be attacked, including the selection of techniques to be employed and the rationale for their choice.

III.7 The *fourth* stage is the determination of the "universe" (i.e., group or area) to be encompassed, and the size and selection of the sample of that universe to be employed.

III.8 The *fifth* stage is the gathering of the data and the processing of it into workable form.

III.9 The *sixth* stage is the interpretation of the data.

III.10 The *seventh* stage is the verification of the interpretation ("conclusions") by either (a) confirming or questioning the results of other studies—for instance, through replication—or (b) confirming or rejecting the original hypothesis.

III.11 The *eighth* and final stage is the presentation of the findings in a report.

III.12 In one form or another, nearly all reputable scientific studies include these eight major stages; and this volume is organized in a sequential examination of all these stages but the last. (This last step involves mechanical questions of reporting, and hence is not functionally germane to the purposes of this volume.) Several remarks relevant to this procedural pattern should, however, be noted: (a) It is purely arbitrary insofar as some textbooks on research methodology list only three, four or five steps; yet an examination of the internal stages of such listed patterns indicates that all eight steps listed above actually are involved. (b) The various stages do not necessarily neatly occur one after the other in all research studies. The experienced scientist may be studying the literature at the same time that he is formulating his hypothesis and is planning both his research design and the selection of his sample. (c) The various steps are not always rigidly determined at the beginning of a study. A well-designed study may permit modifications—of the universe, of the sample, of the data-processing techniques, of the method of presentation, etc.,—to be made during the course of the investigation. (d) All the stages are equally important in contributing to the end results of the study; but not all eight stages involve the same amount of time, cost, manpower or effort.

III.13   This pattern of stages is common to all major scientific studies whether they be in the field of the physical, the biological or the social sciences. Some ambiguity, however, exists in the matter of labeling. Some researchers contend that any kind of serious study properly should be labeled as scientific; others contend that there are several different scientific methods (this problem was discussed in the first chapter); while still others contend that only one specific technique (usually, the laboratory experiment) can properly be designated as a scientific method. Perusal of the literature of the various sciences (and occasionally even of the humanities) clearly illustrates this ambiguity of designation; and too often one finds that even minor studies which employ only two or three out of the total eight steps are seriously purported to meet the requirements of scientific method.

III.14   This ambiguity is a consequence of several different factors. One is the too-common tendency to label any method as scientific simply in order to imply that the method is presumably objective—however unsystematic it may be. Another factor is the popular tendency to bask in the halo prestige of the term science by applying it to any approach, even the obviously specious methods of the quack. Still another factor is the lack of understanding of the analytic function of the scientific approach —a function going beyond the mere encyclopedic approach of data gathering or the "cook book" approach of procedural description to the final stage of answering a specific causal question. These eight stages, therefore, exemplify the total analytic function of science insofar as they provide a common framework which insures an orderly and systematic pattern of procedure which has been found to be most fruitful in terms of the kinds of questions the scientist seeks to answer.

B. *Feasibility of Solution*

III.15   To be suitable for scientific inquiry, a problem first of all must be clearly stated; in short, it must mean the same thing to any intelligent or informed person. Ambiguous questions, equivocal questions, vague or indefinite questions frustrate scientific inquiry right at the start. The question, for example, "Does automation displace workers?" does not specify what types or degrees of automation are referred to, in what kinds of industries, under what conditions, or whether the reference to displacing is intended in the present or future tense (or how far into the future), or whether the term refers to relative or absolute decreases in numbers employed in a given industry, etc. A better question might be "To what extent will the present degree and type of automation utilized in the three largest automobile industries result in a relative change in the present worker-to-unit-produced ratio?" In most cases it is highly desirable for reasons of precision as well as of clarity that a question be formulated in as simple and as limited terms as possible. Complex questions, even like the one offered in the example above, should preferably be broken down into specific and unambiguous component parts.

III.16   Secondly, a proper scientific question must be answerable by available methods of scientific inquiry and by available sources of data. The questions, for example, Will our civilization disappear in time? What would happen if we were to eliminate all social classes? How long would an organism survive if it were completely immune to disease? When will the human race become completely amalgamated biologically?—are not of a type to permit empirical research at this time. Such questions are speculative, hence not factually answerable. Such questions might be regarded as answerable by a nonscientific philosopher, metaphysician, or layman, but never by an empirically-orientated scientist.

III.17   Such speculative questions can, however, at times be approached scientifically if they can be restated in proper form (i.e., in the form of an objectively stated, empirically-testable hypothesis). The question, for example, What would happen if we were to eliminate all social classes? might be answerable—in terms of presently available methods and sources of data—if it were restated in a form such as: What has happened in those cultures that attempted to operate without social classes? Admittedly these two questions are not identical, nor do they entail the same implications. But if they could be accepted as reasonably synonymous, then the second form might suggest an answer to the first.

III.18   Thirdly, a scientific question should be answerable in objective terms. To ask, for example, Does saturation bombing of cities increase or decrease the victims' morale?—is to leave unspecified the meanings of the various key terms. Even if the terms "saturation bombing," "cities," "increase" or "decrease," and "victims" could be objectively defined, it is improbable that the term "morale" could be objectively defined at present to the satisfaction of many persons. The term "morale" possibly might be defined operationally in such terms, for example, as the amount and type of absenteeism from work, the expressed dissatisfactions with civil or military restrictions, the increase or decrease in black-market activities, the increased desertions from civil-defense posts, etc.

III.19   To be answerable in objective terms implies that the questioner and the investigator can agree upon a standard of measurement or upon a definition of the evidence to be sought. It also implies, according to the denotation of *objectivity* (as defined in Chapter I), that the qualities in question to be employed as criteria of measurement or as evidence exist in tangible and impersonalized form, that they have distinctive and denotable features, and that they can be ascertained and described without reference to *subjective* (i.e., self-defined, personally-oriented) terms.

III.20   A special problem often interjected into discussions of scientific inquiry is the question of the scientist's role in relation to value questions. Popularly stated, the question is asked variously as: Can scientists make value judgments? Should scientists study values? How about morals and science? How about a scientific ethic? All such questions can be answered rather simply in terms of the principles stated above. That is, such questions are legitimately scientific if they are posed in clear and unequivocal terms which permit objective answers derived from empirical methods employed within the framework of contemporary scientific knowledge. They are not answerable by a scientist as a scientist if he is expected to take sides in debatable issues simply because of his personal preferences. In short, a scientist approaches moral, ethical or religious questions in exactly the same way and for the same reasons that he approaches any other phenomena to be studied; viz., as objective problems to be submitted to empirical methods which produce verifiable results. If such questions cannot be framed according to his requirements, studied according to his methods, and verified by his techniques, then they are not considered legitimate scientific questions and are rejected as such.

III.21   This means, in practice, that the scientist will avoid as methodologically unanswerable such questions as: Can it be proved that monogamy is superior to polygamy? Isn't it true that Christ died for our sins? Should capital punishment be outlawed? Aren't raw vegetables better for people than cooked ones? His only possible acceptance of such questions as proper subjects for study would be based upon the demonstrability that the questions can and will be rephrased according to the requirements mentioned above. He might, therefore, respond to the first question (about monogamy): "It probably can be proved (or disproved) provided we agree on an objective criterion of superiority; and presuming that we can agree on the definition of the terms monogamy and polygamy." To the second question (about Christ), he could only answer that the phenomenon referred to does not permit investigation by the methods at his command at this time. To the third question (about capital punishment), he might try to find out how many of what types of people believe that capital punishment should be outlawed, or whether capital punishment seems to be an effective deterrent to certain types of crimes; or what types of capital punishment under what conditions? etc. But he could not approach the question as stated because it demands a purely evaluative judgment to be made independently of empirical evidence. The last question (about raw vegetables), probably would be the simplest to investigate scientifically; provided, however, that someone would objectively define the term "better," and would also specify what vegetables, for whom, under what conditions, in what amounts, etc.

III.22   In short, the scientist operates on two levels: one as a scientist in a given field of inquiry, the other as a layman. As a layman he holds values about many things, just as any normal person does; and he justifies such values on simply subjective grounds of preference, just as any layman does. When he speaks as a scientist within his field of knowledge, however, he answers value-laden questions only when such values are objectively defined for him. This point bears emphasis, first, because laymen are often confused by it; and second, because scientists themselves at times violate it. In fact, some of the noteworthy errors in the history of science stem from this very type of violation, but examples can be found in any contemporary period. (One that comes immediately to mind is the controversy between two Nobel-prize scientists over what is basically a moral, not a factual, question: Should extraordinarily devastating bombs be used in future warfare?) The only explanation for assertions made by scientists in relation to nonfactual questions—especially when such questions involve judgments lying outside their established field of competence—seems to be that many technically competent scientists sometimes apparently forego the scientific approach when they are pressed for answers outside their field of competence. Or it may be just as likely that such persons become much more ideologically involved in questions lying outside their field; and therefore, in effect, throw objective caution to the winds when arguing vital social and moral questions. Whatever the reason, such ethical pronouncements, however laudable or understandable, should not be mistaken for objective statements of fact characteristic of most dedicated scientists speaking within the confines of their data.

III.23 In some instances it may not be known whether a given question can be answered at a given time. The limitations of data, methodology, finances, popular or official hindrances to research, etc., may be indeterminable at the time the question is asked. The usual attitude that the scientist assumes in such cases is purely pragmatic (i.e., "Let's see if it can be done").

He may find in the course of his study that he eventually runs into various obstacles: of technical limitations, of material unavailability, of inadequate assistance, of cultural taboos, of inadequate time or finances, or of any of a number of constituent elements needed to complete his research program. Much research, therefore, is in this sense only, purely speculative; and is carried on in the hope that some contribution to knowledge will come out of the effort expended.

III.24 In other instances, the question asked seems to have no immediate or practical value, and is asked solely for the purpose of satisfying curiosity. This is substantially what is involved in so-called *pure* research. The scientist pursuing a problem in pure research may not even be very sure as to just why he is curious to find out whatever it is that interests him. In the long run, however, the various bits of knowledge gained from pure research begin to fall into patterns or into long-existing bins in man's storehouse of facts; and to the scientist operating on the frontiers of knowledge, suggest answers to basic questions. Most of the spectacular discoveries of science such as atomic fission, the germ theory of disease, jet propulsion, radar, radio, or electric power are the practical consequences of knowledge gained earlier from seemingly useless pure research.

III.25 The dichotomous distinction between so-called "pure" and "applied" research is not very meaningful to us here. Science is fundamentally a particular method of achieving reliable knowledge. In pursuit of that achievement, all its knowledge is theoretical—i.e., it is an *interpretation* of reality, not the reality itself. This theoretical interpretation is often arranged in terms of ideal or perfect forms or conditions—e.g., an absolute condition designated as a "vacuum," a perfect synthesis of two or more chemicals in a liquid designated as a "solution," a specific and discrete entity designated as a "cell," or an absolutely round figure designated as a "circle." Such conceived ideal forms are termed *models,* whether referring to simple concepts such as "solutions" or to highly abstract concepts such as "the law of gravity." But such models are only approximations, and hence tentative interpretations of reality. The function of science, therefore, is to endeavor constantly to refine and improve such models so that they may continually approximate more closely the assumed reality of form and matter in terms of increasing and more refined empirical evidence.

C. *Special Problems of the Social Sciences*

III.26   It is sometimes contended that the scientific approach —at least in its more rigorous aspects—is not applicable to social or psychological phenomena. Three basic arguments are generally employed by those skeptical of a rigorous behavioral science: (a) that social and psychological phenomena are inherently subjective rather than objective; (b) that such phenomena are inherently too complex for analysis; and (c) that human behavior is inherently unpredictable due to the operation of a "free will." The intensity and persistence of this contention—one that has been argued even in the highest levels of our Federal government—suggests that these three claims might profitably be examined in some detail.

III.27   The first contention (viz., that a verifiable social science is impossible to achieve because social and psychological phenomena are inherently subjective), distorts the implication of the concept of objectivity. Undoubtedly human behavior—apart from that which is purely biological (e.g., metabolism)—is a consequence of mental or subjective stimuli ("ideas"); but the *behavior itself* may be quite objective. "Love," for example, may be considered a very subjective attitude; but the *expression* or *consequences* of that attitude might quite agreeably be defined operationally by such objective manifestations as various acts of love making. The distinction that should be drawn here is the possible—nay, the necessary—separation of a mental stimulus inducing a behavior from the resultant behavior itself—i.e., the separation of a cause from its effect. The cause of a human behavior may remain subjective insofar as we do not know how to entirely objectify ideas; but this does not necessarily militate against the objectification of the resultant behavior.

III.28 Much human behavior is just as objectively demonstrable as is nonhuman behavior; and much of it can be described and measured in objective or empirical terms. The subjective factors which induce objective human behavior may not at present be ascertainable; but they can often be inferred. That is, they may be inferred from the observed facts just as antecedents of objective phenomena in physical science often are postulated by inference. (After all, who has ever seen, felt, weighed, heard or smelled electricity, force, friction, or space?) Since all knowledge is essentially subjective in the sense that it is derived by individual human beings, the only meaningful question to be raised at this point is whether or not the human behavior in question can be stated in objective terms, preferably according to empirical or operational referents. As will be noted in the following chapter, this problem actually is not an overwhelming one for the competent social scientist.

III.29 The second contention—that a verifiable social science is impossible to achieve because social or psychological problems are too complex for adequate analysis—depends upon the meaning of complexity. Actually, it has not been demonstrated that human behavior, individually or in the aggregate, is inherently either more or less complex than is the behavior (let us say) of an airplane in flight, of a plague of smallpox, of the construction of a skyscraper, or of the management of a factory. Man has demonstrated his ability to solve many apparently complex problems once he has learned to break them down into manageable components. He has failed only when he has tried to solve the whole (complex) problem before he had achieved the answers to most of its separate parts. Complexity is, after all, not an inherent quality of all social phenomena. To the novice, a foreign language may be a frighteningly complex learning situation; but to the native, that same language is so "obviously" simple as to defy explanation.

III.30 Human problems are not necessarily different from nonhuman problems in terms of relative simplicity (or complexity). They probably seem overwhelmingly complex only when (a) they are not clearly understood, (b) they have not been broken down into functionally related component parts, or (c) when they have not been formulated in terms of legitimate scientific questions which permit solution according to the standardized methods herein being described. There is no reason to assume, for example, that the problem involved in maintaining discipline in an army barracks is inherently more complex than is (let us say) the problem of launching a rocket into orbit around the earth and then returning it to a planned base intact.

III.31 The above statements on complexity might profitably be illustrated at this point. In the area of crime, for example, it has been only recently understood that the causes of crime do not lie simply in constitutional factors (either genetic or developmental), nor in climatic or geographic factors, nor in intellectual factors (at least as measured by I.Q. scores), nor in racial or nationality factors, etc. Furthermore, crime was until relatively recent times viewed as one large general problem. But modern data clearly indicate that crime is a many-faceted problem composed of many interrelated parts (e.g., physical factors, environment, educational level, type and degree of law enforcement, social status, family stability, minority status, types of definition of the concept "crime" itself, etc.). Furthermore, it seems highly probable today that crime may not even be a general problem— just as "colds" may not be. It seems highly probable, rather, that different types of crime are in basic and significant ways quite unrelated both in their etiology (i.e., causes or development) and in their treatment and prognosis.

III.32 Finally, the problem of crime historically has been phrased in moral or occasionally even in supernatural terms (rather than in socio-psychological terms); and in such manners or in such terminology that defy objective and empirical analysis (e.g., Why has the devil possessed his soul? Why shall the sins of the father be visited upon the children? Are criminal tendencies inherited? Can it be proved that incarceration breaks a man's spirit?). All of the above examples contain one or more terms which simply prohibit scientific analysis of the type herein being described.

III.33 The third contention—that a verifiable social science is impossible to achieve because human behavior is unpredictable due to the operation of a "free will"—rests upon the specious theological notion that man behaves as a so-called free agent. When examined, this argument breaks down on two counts. First, man is not truly "free" to do as he pleases. He cannot exercise control over many physical and biological factors that both limit and influence his behavior. For example, he can, to only a limited extent, if at all, control his reactions to infections, aging, light, temperature, air pressure, humidity, muscular deficiencies, sensory limitations, responses to drugs or electricity, pain, pressure or shock. He is not free to control (in the sense of preventing), earthquakes, tornadoes, tidal waves, the course of the sun, etc., and other such phenomena which vitally affect him. And he cannot control (in the sense of eliminating, and if he wants to remain alive), the demands made upon him by his biological processes.

III.34 Second, nonhuman phenomena also are often unpredictable (a) when they are studied individually rather than as whole classes, or (b) when the various influences operating upon them are unknown or not measurably ascertained. Which leaves, for example, will fall from which trees in the orchard, at what times, and in what directions is at present unpredictable, but no one imputes a force of "free will" to such events. Nevertheless, the present inability to predict the individual (as contrasted to the group or class) actions of many nonhuman phenomena has not seriously impeded the growth of an elaborate and functional physical and biological science.

III.35 The supposed inhibiting influence of a "free will" upon

the development of a rigorous social science is further *denied by
the very postulates of science* referred to in the previous chapter.
The assumptions that man is part of the natural world, and that
nature is orderly and regular, together can be viewed as a general
"deterministic assumption." This implies that the behavior of
man, however variant or complex, can be understood by the use
of the scientific approach. Like all postulates this collective one
cannot be disproved on factual grounds; rather, it is an assumption employed for the functional purpose of studying behavior.
The argument of "free will," therefore, is simply by-passed on
the pragmatic grounds that to accept it would deny the possibility
of developing a comprehensive social science at all. Lawful determinism, in short, is a functional device employed in the study
of social causation.

III.36   The formulation of a research problem in the social
sciences, therefore, should follow basically the same steps as does
the formulation of a problem in the physical or biological sciences. Bearing in mind the contentions just mentioned made by
the critics of a rigorous social science, it is especially important
that social or psychological problems be formulated in terms of
observable, objective qualities which can be designated in empirical and preferably also in measurable terms. It is also important
that such problems be broken down into their constituent parts
so that they are of manageable size. And it is most important to
remember that—although the behavior of a single unit of a class
of phenomena is not predictable—probability prediction for a
group or class is highly desirable and useful knowledge. When
one admits the fact that many aspects of human behavior (e.g.,
dietary preferences, death rituals, marriage customs, cultural
regularities) both individually and collectively are at present
better understood and therefore more predictable than are many
aspects of physical phenomena (e.g., earthquakes, weather, cos-
mological occurrences, certain fatal diseases)—then the common
notion should soon be dissipated that social or psychological
problems are inherently too difficult to formulate and solve according to empirical scientific method as practiced by competent
researchers.

D. Scope

III.37 Once the research problem has been clearly defined, the next procedural step is the delimitation of the area of investigation. In some cases this step is implicitly included within the context of the question, but in other cases needs to be specified explicitly. In either case it must be made perfectly clear at the outset just where the boundaries of the investigation will lie.

III.38 A scientific problem may range anywhere from a small, specific and temporary inquiry (e.g., How many persons will contract poliomyelitis in March?), to a mammoth field project (e.g., What is the relation between smoking and lung cancer?). Generally speaking the scope will be influenced by one or more of the following factors: (a) the aims and interests of the researcher or his sponsor; (b) the amount of relevant material available; (c) the complexity of the theoretical framework underlying the research design; and (d) the amount of existing knowledge contributing to the solution of the present problem. In any event, the problem should be of manageable size; much research flounders on the reefs of sheer magnitude.

III.39 A simple factual type of problem is an end in itself. It simply seeks to answer a quantitative question in a specific physical circumstance. This type of effort, generally referred to as the "natural history stage" of a science, consists largely of the accumulation of specific factual data which may be employed in a variety of ways. Largely encyclopedic, this type of knowledge forms the factual foundation of all sciences—in fact, of all verified knowledge. Many sciences, even today, are composed largely of nothing much more than large masses of discrete data. Examples of such rudimentary sciences would be physical anthropology, seismology, linguistics and political science. In most cases, however, a given research problem is meaningful only when viewed as a related part of a larger general problem. At the outset, then, it is important to orient the problem in terms of both its purpose and its interrelation.

III.40 In terms of purpose a problem is generally designated as either or both of two general types: (a) It may be purely descriptive; for example: Who is infected? Where can coal be found? When did the chrysalis stage begin to form? What kind of bacteria are these? (b) It may be analytical; for example: What is the interrelation between clouds and rain; or between age and disease; or between income and I.Q. scores; or between humidity and corrosion? A third type, the exploratory study, is essentially one of these two general types. It differs only in the sense of being less rigorously structured. Pending a more detailed consideration of this question of designation later, it should be mentioned at this point that the terms "descriptive" and "analytical" are employed here only for convenience of general usage. Actually, the two processes, description and analysis, are in most cases so closely interrelated that the distinction between them is artificial rather than real. The two terms, therefore, are temporarily employed as a functional convenience in order to simplify their treatment.

III.41   In terms of interrelation, a research problem should be oriented to the larger field of theory and knowledge of which it is a part. To proceed—except in a simple census-taking type of problem—independently of the larger field is to invite error or wasted effort. It may be obvious in some cases how a given problem is logically related to a larger "parent" field of theory or knowledge; but if this relationship is not obvious, then the research design should indicate that relationship.

III.42   Thus it should be clearly stated how a particular study is related to other studies of the same type; whether the study is a pioneer without any antecedent examples of style, method or purposes; or whether it is simply a replication of others, being done for the purpose of confirming their findings. It should be clearly stated whether the study assumes that the findings of like studies are accepted without question, or whether they will be ignored, disputed, or possibly even disavowed. The orientation of a problem indicates to the reader that the researcher has a specific functional purpose for studying what he does and in the manner that he does. In the chapter to follow, these principles of organization will be related to the next major step of scientific research: the formulation of clear and functional concepts and constructs.

E. *General Principles*

III.43   The heart of any scientific problem is its correct formulation. In fact, it has often been said that the major problem of science is that of asking the right questions. To that end, some general principles of problem formulation characteristic of good methodology will be illustrated below.

III.44   (*1*) *Be sure that the problem exists.* The patient who persistently returns to his physician week after week because of an acute abdominal pain might be suffering from a variety of real ailments ranging from ulcers to cancer. The cautious diagnostician tests as far as is practicable for all suspected causes of the pain. But he should also be aware of the fact that (a) the pain might be imaginary rather than real, (b) it might not actually be occurring in the abdomen, or (c) it might exist but not from internal causes. Only the inexperienced physician would assume that the patient's analysis is unquestionably correct. To pursue this illustration, then, the problem might actually exist as defined by either the patient or the physician; but it might actually not exist at all as defined by the patient if it is of psychosomatic origin—or if the patient is a hypochondriac. Therefore, assurance of the existence of a problem requires a clear definition of its attributes as well as a clear delineation of its area of existence. It would be futile, for example, to attempt to solve the problem of excessive alcoholism if the subjects involved (whose help would be needed in the solution) had not defined their drinking habits as constituting a problem in the first place. It would be just as futile to attempt to solve the problem of divorce if in reality the real problem is not divorce itself (which is now generally regarded as just a symptom) but, rather, is a complex of such related factors as uncoordinated social change, the decline in traditional authoritarian patterns, changes in traditional sex roles, economic factors, etc.

III.45   (*2*) *Learn as much as possible about the problem to be solved.* Too often an investigator attempts to solve a problem before he understands clearly either its component parts or its distinctive and recognizable attributes. It would obviously be foolish to seek the causes of headaches, for example, before one had a clear definition of what a headache is, how it can be recognized, how it can be differentiated from other aches, or how it can be measured and recorded. A basic knowledge of physiology and neurology certainly would be called for, as well as a knowledge of the factors which possibly precipitate a headache. Yet research in some fields does often proceed without this basic general background of contributory knowledge; so it is not surprising that poor preparation often produces nothing more than poor results. (Consider, for example, the obvious case of searching for solutions to the problem of war before one knows even the basic and common causes of war—if such actually exist; let alone if one cannot distinguish a "war" from a "police action.")

III.46   (*3*) *Employ the most feasible or most efficient methods of solution or both.* This principle usually requires a judicious consideration of cost, time, manpower available, etc., before a problem can be formulated into a research design. For some purposes, simple but not too efficient formulations may serve the needs at hand; while in other cases, the problem demands the most rigid and exacting formulation possible. It would be inefficient, for example, to spend large sums for complicated research designs to solve a problem of miniscule dimensions or significance; but it would be just as foolish to attempt to answer large complex questions by the utilization of primitive and limited methods. The trite statement that the method should fit the problem only serves to beg the question; for in many cases, several alternative formulations or methods might fit the problem equally well—depending upon the depth and breadth desired in the solution. Whatever the actual answer, the competent scientist should, when he formulates his problem, have at his command a thorough knowledge not only of various methodologies, but also of the relative efficacy, cost, time, labor, etc., involved in their employment. Experienced researchers often formulate their problems according to different designs, or employ differing methods of solution in order to check one against the other; but generally speaking, most established research problems suggest in their definition which methods are most efficacious for solution.

III.47 (*4*) *Consider alternate or substitute formulations in case the original one is not feasible.* In spite of the best possible planning, it sometimes happens that the original formulation of a problem cannot feasibly be executed in terms of a specific research design. In other cases, the original formulation loses its significance due to the passing of time, to changes of interests, to changes in the subject to be studied, to decreased financial support, etc. The prudent scientist, therefore, keeps in mind alternate ways of reaching his objective. This may mean a lowering of the sights of achievement, or it may mean the taking of a more indirect road in order to reach an approximation of the original goal. Like the careful planner and strategist that he should be, the experienced scientist is always ready to make changes in formulation or procedure insofar as the eventual objective is not discarded. This means, in effect, that he needs to know exactly how many and what kinds of changes in the original formulation are possible without seriously altering the essence of the problem. This is a knowledge acquired only after thorough familiarity both with one's field of interest and with the whole scope of scientific methodology.

III.48 (*5*) *Check for recognition of the phenomena and also for false recognition.* The practice of using so-called "positive controls" in industry is well established. This practice consists in placing in a group or series a faulty or atypical unit in order to check the functioning of detection instruments or of inspectors. In short, this principle means that a problem cannot be solved correctly if the phenomena are not clearly identifiable as such, or if the techniques of analysis or the instruments employed are not both adequate and reliable. The common example of seeing ghosts when ghosts do not exist, of "not seeing the forest for the trees," of "proving" the existence of gold by exhibiting a handful of iron pyrites ("fool's gold"), or of ignoring the "black mud" of a rich but unrecognized oil field—all attest to the fact that a problem cannot be correctly solved if the phenomena inherent in it are not recognized or are falsely recognized. The use of a placebo (i.e., an ineffective or inert substance, such as the "pink pill" employed in some experiments) is standard operating procedure in many research designs to test the responses of subjects to real or imagined effects. Just as well known are the classical demonstrations of suggestibility employed by psychologists or by lawyers: viz., that most laymen can easily be convinced that they see, hear, feel, taste or smell things which do not exist in fact— or, conversely, that they do not perceive even obvious things under conditions of negative suggestibility. Therefore, an adequate problem formulation will include so-called positive and negative recognition checks in order to achieve accuracy.

III.49   (6) *Formulate the problem systematically.* Since the second attribute of science—exactness was the first—is systematic procedure, it is almost redundant to state at this point that a problem should not be formulated haphazardly nor should its design include haphazard procedures. Yet the history of science as well as a perusal of some contemporary scientific literature would attest to the fact that many significant problems remained unsolved for an unnecessarily long time due to formulations that lacked any recognizable order. To anticipate an extensive discussion of research design in Chapter V, it might be useful to point out here that a systematically constructed study design is impossible to achieve if the problem has not been formulated in a systematic manner. Though many problem formulations require changes during the course of the research procedures, such changes can be logically defended and maximally exploited only when they can be referred to a problem which originally was formulated systematically. To proceed haphazardly may be an exciting adventure when visiting a strange city or when buying a new hat; but in science the achievement of meaningful results is a consequence only of systematic formulations coupled with exact methodology.

III.50   (7) *Do not try to solve complex problems by simple formulations.* There are instances when a complex problem can be broken down into separate and discrete parts; in such cases, each separate sub-problem can be formulated into a meaningful research design by itself, and the combined results therefrom can be coordinated, amalgamated or synthesized to answer the original complex problem. In most cases, however, the reverse is true. That is, a complex problem by definition means a problem whose many parts are intimately and inextricably interrelated. If the formulation ignores this basic fact, and then proceeds to try to answer only the separated parts, only insignificant conclusions can result. Simple examples of the violation of this principle are found most often in the social sciences, where discrete findings derived from simple formulations are frequently offered as the answer to complex problems—for example: What is the influence of formal education upon marital happiness? The answer to this question cannot be achieved according to such a simple formulation; for it is obvious to any informed student of the subject that formal education, in relation to marital happiness, does not operate independently of such interrelated factors as income, occupation, number of children, race, religion, nationality, region, length of marriage, etc. Only if all these other interrelated influences had been determined and could have been held constant—a task to frighten even the most audacious statistician—could the problem then perhaps have been formulated as stated originally.

III.51 *(8) Be aware of the possibility that the problem formulation may influence the phenomena being studied.* It is an elementary fact known to any physiologist or psychologist that human subjects, knowing they are being studied, often react differently than they would under normal conditions. It is also an elementary fact to a physical scientist that many samples or specimens react differently in the laboratory or under test conditions than they do in their natural environment. The careful researcher, therefore, formulates his problem—and later constructs his research design—in such ways as to avoid, counteract, or control any possible effect that the testing situation may have upon his phenomena. Although this problem of subject reaction occurs in all fields of science, it is particularly when dealing with human subjects that this principle has its greatest applicability. Experience is an invaluable teacher at this point; and each field of research has developed different techniques to solve this particular problem. As a final resort, specific checks for "test sensitivity" on the part of the subjects, or careful comparison of test subjects with untested samples, are the best insurance against the possible violation of this principle.

III.52 In summary, then, it should become increasingly apparent why the formulation of a research problem is such a critical step in the total scientific effort. The need to define the problem clearly is imperative to the achievement of any degree of success; and the shoals and reefs of human endeavor are littered with the wreckage of poorly posed problems of research. Particularly in some areas of the social sciences—where the feasibility of employing rigorous scientific methodology is still subject to serious doubt—adherence to the established principles and methods which have proved so fruitful in other fields is long overdue. Two prevalent practices are particularly significant here: (a) that of simply gathering data for their own sake and without methodological concern for their employment in a substantial research design, and (b) that of structuring data in a wide variety of designs of dubious quality in the sanguine hope that some significant conclusions might thereby be derived. As will be noted in the discussion of analytic designs later, in chapter IX, the whole structure of empirical science stands or falls upon the foundation of clearly formulated questions logically arranged into a meaningful research design.

CHAPTER IV

The Role of Concepts

A. CONCEPT DEFINITION

B. UNITS OF STUDY

75

A. Concept Definition

IV.1   Since science strives to achieve accuracy, every field of scientific endeavor develops a continuously refined set of concepts which, to the initiated, mean the same thing at all times under stated conditions. Thus it is imperative at the outset of any research effort to define clearly every *concept* (i.e., an idea, or a generalized idea of a class of objects), or *construct* (i.e., an idea expressing an orderly arrangement of concepts into a single whole) that will be employed. (The term "construct" is often employed to refer to abstract or purely synthetic formulations having no counterpart in observable reality—e.g., "force," "symbiosis," "status," "value." Treating such abstractions as though they do exist in tangible form—e.g., visualizing the "status hierarchy" as a real pyramid of roles, or the earth's core as a magnet inducing gravity—is termed "reification." To reify is a common error made by those who think primitively.)

IV.2   The problem of accurate definition is of fundamental importance in science. One of the basic rules of definition is that a definition can be neither true nor false—i.e., it is not a factual proposition. A definition is simply an explicit declarative statement or resolution; it is a contention or an agreement that a given term will refer to a specific object. One may question the intelligibility or the usefulness of a given definition, but he cannot logically test its truth; for its "truth" is established by declaration—i.e., it is what the definer says it is.

IV.3   In order to promote clarity and precision, several basic rules of definition are consistently followed by all competent scientists in accordance with accepted principles of logic. Stated briefly they are, first, that a definition must denote the unique or distinctive qualities of that which is being defined. The term employed must be the symbolic equivalent of the thing it stands for; it must be applicable to every instance of that thing and to nothing else. In other words, it must be inclusive of all things denoted by it and yet exclusive of all things not denoted by it. This quality of precision and distinctiveness is of particular significance when building a *taxonomic* system (i.e., a system of classification of objects wherein each class bears some logical relationship to every other class); for the usefulness of any taxonomy is directly related to the precision of its various classes of objects. It would be of little value, for example to classify certain objects as "foods" if such things could not be further defined more precisely and exclusively in terms of specific qualities (e.g., their use by different species, their caloric content, their commercial value, their nutritive value or their scarcity). The development of a comprehensive, precise and functional set of classifications is a significant index to the degree of maturity of any particular science.

IV.4 The second rule of definition states that a definition must not be circular—i.e., it must not contain within itself either directly or indirectly any part of the thing being defined. The error of *tautology* (i.e., of defining a thing by itself—e.g., "A man is a person having masculine qualities"), is a feature of vague exposition such as is commonly found in small, cheap dictionaries. The difficulty here, however, lies in the fact that very few terms have true equivalents in any language; so the problem of tautology is sometimes difficult to resolve. To say, for example, that a boat is "A small, open vessel or water craft," may seem clear enough until one checks the definition of the term "vessel," only to find that it refers to "A craft for travelling on water," while a "craft" refers to "a boat, ship or aircraft." This second rule, therefore, refers mainly to gross or obvious errors of tautology such as might be the case when one defines a "crowd" as "a group lacking organization." A contemporary example of a popular tautology is the case of defining the aged in our society as "senior citizens." Later, under a discussion of common errors of conceptualization, this type of error will be analyzed in more detail.

IV.5 The third rule of definition states that a definition should not be stated negatively when it can be stated positively. This rule, however, is not as binding as are the others; for in some instances the exclusiveness feature of a good definition demands that it be made clear what a term does not stand for. (For an example, see how the term "natural" was defined in

Chapter II.24.) Furthermore, some concepts are essentially negative in character. A drunk, for example, is essentially a person who is not sober, and a loser is simply one who is not a winner. Whenever possible, however, clarity is enhanced by the employment of positive rather than of negative denotations. In cases of doubtful interpretations, both the positive and negative features of a term may be utilized to insure clarity.

IV.6 The fourth rule of definition states that a definition should be expressed in clear and unequivocal terms, not in obscure or figurative terminology. The problem here, however, is often that of common agreement and understanding of terms. What might appear as clear and unequivocal to one person might appear vague, amorphous or obscure to another. As a case in point the term "disease" might prove illustrative. To a physiologist the term may have a very precise meaning; but the same term in the hands of a social reformer (who might speak of a "diseased society") might well be vague, amorphous and highly general or ambiguous. Scientists themselves, for example, often employ terms which may appear vague or obscure to the uninformed person, but which within their group are quite clearly and meaningfully perceived. (Examples of such terms might be "meson track," "cartilaginous tissue," "temperament," "status index.")

IV.7 While on this subject of definition it perhaps would be useful to consider also a basic error often made in the employment of concepts. The error, essentially, is that of assuming that changing or manipulating a word (symbol) changes the thing for which it stands; or, in other words, that things can be changed simply by changing their names. The converse of this error is that of assuming that the same name necessarily implies the same meaning, or that things can be made the same simply by giving them the same names. Examples of this general class of errors in definition, conceptualization or symbolism come easily to mind. Though it may be quite true that "A rose by any other name would smell as sweet," it is highly dubious that a belligerent military program is any the less offensively oriented simply because it is called a "defense program," or that deceit in business can be denied simply by calling it "shrewdness." Since there are so many types of communicative errors possible in any discourse, and since understanding and avoiding such errors is fundamental to scientific competence, an exposition of some major conceptual errors is presented later in this chapter.

IV.8 Whenever possible, concepts and constructs should be defined either (a) objectively or (b) operationally—i.e., they should be defined (a) in terms of empirically verifiable and standardized referents (such as rulers, thermometers, scales, etc.), which leave little room for dispute among competent observers; or they should be defined (b) in terms of specific operations, behaviors, processes or effects which likewise leave little room for serious dispute.

IV.9 Thus terms such as "beauty," "good," "rich," "intelligent," "hot," "wide," "dark," etc., are useless for research purposes until they are transposed into objective or operational referents. "Beauty," for example, might be defined objectively as "A score between seven and ten on the Smith Scale of Beauty Ranking," or it might be defined operationally as "Evidenced by winning a contract to pose for perfume advertisements." "Good," for example, might be defined operationally as "A good student is one who is never cited for disciplinary infractions." "Rich," of course, might easily be defined objectively in terms of gross income, or savings, or property owned. "Intelligent" might be defined objectively as an I.Q. score above a certain designated point; or it might be defined operationally as "Solving the maze within thirty seconds." "Hot," of course, can be translated into certain objective units of Centigrade temperature, or in terms of volume of heat by "BTU's" (British Thermal Units); just as "wide" can be defined objectively in terms of inches, yards or meters; while "dark" might be defined objectively as "Any condition less than ten candle-power of measured light."

IV.10 A particular problem of definition arises in the employment of measuring instruments. The standard practice is to define an instrument in terms of two qualities: validity and reliability. An instrument (a scale, ruler, balance, meter, questionnaire, attitude test, etc.) is said to be *valid* when it measures that which it is purported to measure. It is said to be *reliable* when it gives consistent results under comparable conditions.

These two features are not necessarily related. The validity of an instrument is generally established by consensual definition—for example, the agreement that a valid measure of foot candles (of light) or of viscosity (of a liquid) shall be that as determined by such a body as the American Standards Association or by the Society of Automotive Engineers. If the definition is either ambiguous or not agreed upon, however, it is impossible to determine the validity of an instrument. In the social sciences, for example, the lack of a clear agreement on the essential objective attributes of the term "social class" would be a case in point—making, in this case, the development of a valid measure of social class impossible.

IV.11 A related aspect of the problem of determining the validity of an instrument arises in those cases where the logical implications of a definition can be seriously questioned. A case in point again can be drawn from the behavioral sciences in the matter of defining, say, intelligence. To argue that a test validly measures intelligence simply because one has declared that the items *are* measures of intelligence—of a particular sort, at least—is to raise the question of the requirements of a good definition. Unless it has been established that the items tested are the unique and inherently distinctive qualities of what is normally connoted by the term "intelligence," it is dubious that one is actually measuring what is logically implied in the connotation of the term. This same point could be illustrated in the case of such terms as "adjustment," "interaction," "status," "institution." In short, an instrument is said to be valid when it measures those qualities or attributes clearly and objectively defined according to a logically defensible empirical connotation.

IV.12 The relative merits of denotative, connotative and operational definitions, as employed in science, comprise a lengthy topic of debate. The major arguments of that debate, however, can be summarized as follows: A *denotative* definition has a nonverbal referent—i.e., it "points to" the object (often by use of an illustration). A *connotative* definition, on the other hand, implies or describes—usually by listing the attributes or features of the object—that which the term names (i.e., the referent or object). An *operational* "definition," as discussed earlier, describes or prescribes the steps or procedures required to carry out the idea in question. Each of these types of definition has its particular strengths and weaknesses when employed in science.

IV.13   The strength of a denotative "definition" is its essentially nonverbal character. Any normal person can perceive a "horse," a "motor," a "house," or a "microscope," when these objects are indicated by various denotative definitions. Suppose, however, that not all horses, motors, houses or microscopes look, act, or feel alike? How is one then to perceive their common qualities? Herein lies the weakness of a denotative definition. Unless a class of objects has a common, easily perceivable set of attributes—which many classes of phenomena do not—it becomes quite difficult to designate such classes by a simple denotative term. Particularly in the employment of constructs—which by definition are synthetic abstractions having no physical or behavioral referents—a denotative definition is practically useless.

IV.14   The strength of a connotative definition—the type most often employed in dictionaries—lies in its synthesis of the inherent and unique qualities of an object in terms already understood. A "horse," for example, may be defined connotatively as a "four-legged animal about four to six feet tall, having a long bushy tail," etc. Furthermore, abstractions (e.g., "honor," "beauty," "integration," "parabolic") can only be defined connotatively. The weakness of such definitions, however, should be quite obvious. Suppose a person does not understand the referents themselves? How, for example, would one connotatively explain to a primitive person the essential meaning of such terms as "expansion coefficient," "osmosis," "trauma," or "adjustment" if the person did not already possess some related or synonymic referents to which such terms allude? Defining a concept in terms of other, already understood concepts assumes that one understands the "other" terms. As anyone well knows from the experience of consulting a dictionary, a connotative definition is often nothing more than an exercise in increasing confusion.

IV.15   To reiterate, then, it is often advisable and sometimes even necessary in science to define concepts both denotatively and connotatively. In such cases the combined definition practically becomes *operational*—i.e., the concept (but not the construct) is defined in terms of the procedural steps involved (such as "cooking," "heating," "measuring," "testing"). (This procedure has been discussed earlier.) In spite of all its obvious virtues, however, the essential weakness of an operational definition should be mentioned, to wit: there is always the possibility that various users will not agree to define an operational concept in a similar manner. Clay, a bench, an apple pie, or even a specific painting may all be produced by varying operations; and one set of procedures to create, interpret or construct such objects may be just as legitimate (and even useful) as another. In short, an operational definition often lacks the feature of exclusiveness— a feature basic to all good (i.e., precise) definitions employed in science. Other weaknesses of operational definitions—though they interest the logician of science—need not concern us at this point.

IV.16   There are many occasions in science when a concept or construct cannot very well be "defined" in either empirical or operational terms. This occurs in the case of theoretical definitions serving as explications of large classes of phenomena. A general theory of energy, for example, or of mutation, or of social change, or of "homeostasis" (i.e., the tendency of an organism to seek a balance between its tensions and its environment), may refer to abstractions which exist only conceptually but not empirically—and hence not operationally. True, the abstract concept in such cases may be further defined and hence related to specific definitional attributes which can then be related to empirical qualities; but the original definition remains substantially a theoretical abstraction. After all, no one can see, hear, touch, smell or taste "force," "energy," "adjustment," "leadership" or "morale." He can only experience the empirical consequences of such abstract notions when such consequences have been designated by definition.

B. Units of Study

IV.17   Clear definitions have their first applicability when a study delineates its specific phenomena of interest. In some instances the phenomena may be gross classes (e.g., the world's human population, all living bacteria, all forms of heat, the whole planetary system). In most cases, however, a study is concerned only with specific sub-classes of phenomena; hence it must designate such sub-classes in terms of relatively precise units of reference. Such units are the specific features of the phenomena that interest the researcher (e.g., the "income" of a population, the "mobility" of a social group, the "tensile strength" of a metal, or the "viability" of an organism). In order to increase accuracy and precision, therefore, a satisfactory unit of scientific analysis should possess at least five clarifying attributes: appropriateness, clarity, measurability, comparability and reproducibility. The following discussion will indicate the role of these attributes.

IV.18   The first requirement of an accurate unit of study is its *appropriateness;* that is, the unit selected must focus attention upon the essential object of study. Thus an analysis of income differences would need to prescribe whether "income" shall mean the gross salary and earnings, or the take-home pay after taxes have been deducted, or the take-home pay after taxes and contributions and union dues have been deducted, etc. A study of comparative birth rates would need to specify who is being compared: the total population of two cities, or all the females of two cities, or all the married females of two cities, or all the marriageable females of two cities, etc. In general, both a clearly defined question and experience in the particular field of study (besides a clearly formulated hypothesis) are necessary before the appropriateness of the units can be determined. But in the first as well as in the final analysis, the appropriateness of a unit will be determined by its role in the total study design—i.e., by whether or not it fulfills the needs of the study and accurately conveys what is intended.

IV.19   The second requirement of an accurate unit of study is *clarity*. Essentially this is a problem of precise and unambiguous definition. In speaking of "colds," for example, can the unit be precisely defined so that it means the same thing to all students of the subject? The same can be asked of such units as "crime" (All types? Just those called to the attention of the police? Both major crimes and misdemeanors? etc.), or "wars" (Only those officially declared? What about so-called insurrections or revolutions? What about so-called "crime-wars"? And what is a "cold war"?), or even "religiosity" (Going to church? If so, how often? Organized into sects and denominations only? Any differences between believers and doers? etc.). This problem of clarity is particularly acute when a study attempts to employ such subjective, relative or abstract units as "peace," "good government," "economic recession," "morality," etc. In the last analysis clarity is a matter of the degree of specificity, on the one hand, but also of the character of the concept on the other. Some objects inherently permit clear and discrete definition (e.g., cast iron, teeth, pregnancy, death), while others are inherently obtuse or indiscrete (e.g., windy, will-power, illness, reverence). Wherever possible, therefore, clarity should be viewed not as an absolute necessity without which a study could not proceed but, rather, as an ideal goal constantly to be striven for even if never completely achieved.

IV.20   The third requirement of an accurate unit of study is its *measurability*. Essentially this means that one should strive constantly to devise units which permit quantification and therefore mathematical manipulation. Admittedly this may not always be possible; but since mathematics is the most precise, logically consistent, universal and standardized language of science, mathematical measurement is the optimal tool of all scientific endeavor. In any event a unit of study is improved to the extent that it can be defined in measurable terms.

IV.21   The fourth requirement of an accurate unit of study is its *comparability*. This means essentially that the units to be studied and compared should be of a like order. (This again is a definitional, hence taxonomic problem.) Divorces, for example, are hardly comparable (as any student of the subject knows), because of the differences in divorce laws among various states and countries. Nor are such general phenomena as crime, drug addiction, infant mortality, unemployment or migration comparable. In all these cases, although the units are the same, the ways of determining them vary widely. Thus it is necessary that the researcher demonstrate at the outset the comparability of his units. If he is to gather examples of religious behavior, or of drunkenness, or of wife beating, or of poverty, etc., he must be sure that his units refer to the same phenomena wherever and however derived.

IV.22   The fifth requirement of an accurate unit of study is its *reproducibility*. Since science is concerned with general and not with unique phenomena, any study which employs units that cannot be reproduced defies verification. The study of history quite clearly illustrates this particular deficiency of an accurate unit of study. Since most historical events are presumably unique, they cannot be reproduced; so it becomes impossible to restructure them in a manner permitting restudy or verification. Once again, however, it might be well to point out that this requirement, like that of clarity, is neglected more often than is apparently necessary—i.e., it is highly questionable that many presumably unique phenomena actually are unique. (A highly publicized study of human sex behavior, for example, contended that even such intimate—and therefore presumably individualistic or unique—behaviors as sexual relations occur with much greater consistency and similarity than is commonly assumed to be the case.) Final determination of reproducibility or uniqueness is oftentimes a matter of opinion rather than of established fact; but the competent scientist deals only with demonstrably reproducible phenomena and employs them in such manners or designs that both the phenomena and the design can be reproduced by other investigators interested in verifying his conclusions. A major difference between the so-called natural sciences (i.e., physics, astronomy, chemistry and biology) and the social sciences is this very attribute of reproducibility. "Replication" studies are rare in the social sciences though very common in the natural sciences.

C. *Common Errors in the Employment of Concepts*

IV.23   Though science seeks to minimize errors, no method can be more accurate than the person who employs it. So, since scientific method is an instrument of human reasoning, the reasoning processes themselves should be checked for possible sources of error just as are all other tools or instruments. From the first rumination about a problem to the final statement of the conclusions, the scientist must carefully check each step of the reasoning processes underlying his procedures in order to minimize error. But human reasoning does not operate as precisely and as predictably as does a machine. Although scientific method attempts to achieve maximum precision within its procedural framework, even a single error of reasoning can spoil and vitiate the most precise procedures devised. For purposes of convenient reference, therefore, the most common (but sometimes subtle and often insidious) errors of thinking will be explained and illustrated in the section to follow.

IV.24 It should be borne in mind when considering these errors, however, that the classifications are purely arbitrary for purposes of simplicity. As any logician knows, many of these classes overlap, many are derivations of others, and many are refinements of others. The listing to follow, therefore, can be justified only on the grounds of convenience, since it admittedly violates the purism of formal logic.

IV.25 The reader might wonder why the following listing of logical errors was not appended to the previous discussion of reasoning processes presented in Chapter II. The rationale is simply functional. Most of these errors are a consequence of abuses or distortions of the various principles previously discussed in Chapter II. Some of them, however, are specifically related to the problem of accurate and clear definition itself (as just discussed), and therefore are only indirectly related to the deductive processes discussed earlier—or, for that matter, to the inductive processes to be discussed later in Chapter VIII. Their placement here is justified only on grounds of convenience, not of logical order or relationship. Furthermore, these examples are brought together at this point because many of them combine two or more types of distortions within themselves. Since this is not a text-book of logic, the interested reader can find the specific relation between each of these abuses and their logical referent by referring to Chapters II and VIII, as well as to the previous section of the present chapter.

IV.26   (*1*)*Tabloid thinking:* In common usage, tabloid thinking refers to the practice of employing familiar concepts which
connote a general idea belonging to a particular framework. Such
concepts may be either *omnibus* types (i.e., terms which connote
either a positive or negative attribute, e.g., motherhood, beauty,
honesty, intelligence, on the one hand, or filth, ugliness, lechery,
immorality, on the other), or *stereotypes* (i.e., terms which connote generalized classes in their entirety while ignoring the actual
and significant differences existing within those classes, e.g., Negro
athletes, Italian musicians, travelling salesmen, or women drivers).
Such omnibus types or stereotypes may also include music (e.g.,
some keys, chords, tempos or arrangements suggest the somber
mood of mourning, the joyous mood of a country dance, the ex-
hilarating mood of a military parade, or even the soporific mood
of a lovers' conversation); colors (e.g., certain colors are generally
regarded as gay, others as somber, others as violent, and still
others as restful); signs or symbols (e.g., the hitch-hiker's crooked
thumb, various roadside symbols, the cross, the flag, the star of
the policeman); or even personality types (e.g., sketches, cartoons,
paintings or photographs of the presumably typical jolly fat man,
the absent-minded college professor, the old maid school ma'arm,
or the curvaciously alluring movie queen). Their effect, regardless of type, is to induce a predetermined reaction which negates
objectivity. Advertising, political cartoons, publicity photographs,
many editorials, movies and drama, and much nationalistic history and civics clearly illustrate this device of tabloid thinking.
    IV.27   Tabloid thinking is not confined to the layman, how-
ever, for the history of science also clearly illustrates its insidious
presence even in some outstanding works. Within recent memory
one has only to recall the examples of the physicists' notion of
force or matter (i.e., that the two were distinctly separate entities), of the biologists' notion of viruses (i.e., that they were
simply ultra-microscopic organisms), of the psychologists' notions of the relation between sex and intelligence (i.e., that some
aptitudes were biologically sex linked), or of the economists'
oversimplified distinctions between differing economic systems
(e.g., between so-called capitalism and so-called socialism).

IV.28   In each of these (and many other) cases the concept alluded to implied a class having certain common and specific qualities; whereas in reality the differences inherent within those classes often were greater than the presumed likenesses. Furthermore, even methodological concepts may be stereotyped in science. A glance at many presently popular textbooks would illustrate this type of error clearly; as, for example, when authors refer to "the" experimental method (actually there are several different forms), "abnormal" behavior (there is no clear distinction between so-called normal and abnormal behavior), "intelligence" (there are apparently many different and possibly unrelated kinds), or social "institutions" (there are literally dozens of differing definitions of this ubiquitous term in the social sciences). Suffice it to say that this large class of errors can be avoided only by strict adherence to the principles of conceptualization discussed earlier.

IV.29   (2) *Preference for the familiar:* This type of error results from the commonly observable fact that most persons prefer the familiar to the strange. Whether the error takes the form of belief through simple and sheer repetition, or whether it takes the form of a slogan or other cliché, most persons gullibly accept as true that which is merely familiar. Examples are numerous in everyday life, especially when this error is combined with some form of tabloidism (e.g., Truth will out; Blood will tell; With men who know tobaccos best, it's El Ropos ten to one; Vanity, thy name is woman). Although serving as the backbone of commercial advertising and political propaganda, sheer repetition abounds also in the history books of every nation and many examples from the record of mankind attest to its persuasiveness.

IV.30   The error of preference for the familiar can have as deleterious an effect in science as it can when employed in everyday life. Among the more common examples may be cited the belief that the earth was flat, that the sun revolved around the earth, that foreigners are inferior and inherently menacing, or that miracles do occur. More recent examples of this error in science might include the belief that passage through the sound barrier would result in the disintegration of an airship, that a stopped heart could not be revived, that insanity could be caused by the menopause, or that social eminence is a consequence of superior genetic attributes. In each of these cases, the only validity contained was that of commonality of belief, not of empirical demonstrability. Yet the very persistence of such notions in the absence of any objective verification attests to the power of this type of erroneous thinking. In science as well as in everyday life, there seems to be a measure of truth implied in the statement "I believe what I know, and I know what I believe." This is simply another way of implying " I know what I like, and I like what I know."

IV.31 (3) *Incompleteness:* Much erroneous thinking stems from the fact that an argument may employ either (a) *selected instances* (to the exclusion of other relevant instances), or (b) *card stacking*—i.e., amassing only supportive arguments while ignoring destructive ones. In the former case, proof by selected instances implies that those instances selected are representative of all cases of the same kind; while the latter implies that sheer weight of numbers is synonymous with proof. In either case, this type of error violates several logical principles, particularly that of *non sequitur* (viz., the conclusion does not follow).

IV.32 Examples of this class of errors in everyday life come quickly to mind. Proof by selected instances might be illustrated by such statements as: "You can't tell me about Irishmen, I knew one once"; "A woman double-crossed me once, I'll never trust one again"; "Once a thief, always a thief"; "He murdered once, he'll commit murder again if we don't lock him up for life." Proof by card stacking can be illustrated by such statements as: "There are forty-nine reasons why you should buy Madame X's rejuvenator"; "Old Socko contains seventeen different distilling secrets"; "Fifty-million more filters, twenty-one new improvements, and eighty-four different tobaccos more than any other cigarette"; "Ninety-nine million bottles full of eighteen rare herbs plus sixty-two laboratory checks after ninety-one different inspections prove its superiority over all other snake oils."

IV.33 In science these two types of errors of incompleteness might be illustrated by many examples which employ, in the case of proof by selected instances, very small or atypical samples and then imply that those samples are representative of the larger but undetermined whole. Examples of this class might include the following: (a) "Attributes of happily married couples"—implying that happily married couples exhibit common distinguishing features of a diagnostic or analytically significant sort, which, as of this writing, they don't. (b) "Traits associated with delinquent tendencies"—implying that potential delinquents exhibit common behavioral attributes, which they evidentially don't. (c) "Characteristic responses of college freshman"—implying that college freshmen can be characterized in terms of significant common responses, which they can't. In the case of proof by card-stacking in science, examples might include the voluminous though highly selected data purporting: (1) to demonstrate objectively the operation of extrasensory perception; (2) to delineate objectively the specific physical traits ("stigmata") which assumedly characterize all criminals; (3) to "explain" personality in terms of body types ("endomorphs," "mesomorphs," etc.); or (4) to establish factually that anti-social traits are a consequence of specific racial heredity. As will be noted later in a discussion of verification, it is often difficult to differentiate this device of card stacking from proof by consensus. For this reason, modern science tends to regard proof simply as a qualified expression of probability (e.g., "In nine cases out of ten, this tends to occur, . ."), rather than as a categorical expression of fact.

IV.34 (4) *Irrelevancy:* Part of the large class of errors of *non sequitur,* this class employs some form of argument or proof which is logically unrelated to the proposition. It may take the form of ridiculing the opponent and inferentially thereby his views (e.g., "Don't vote for Jones for mayor, he knits, sews and plays tic-tac-toe"); or of misdirected argument (what is commonly referred to as the "red-herring" device of dragging misleading arguments across the path of debate). Thus, for example, the claim that X is superior to Y because X is twice as fast as Y implies (a) that speed is a desirable quality (which it may not be); or granting that it is, implies therefore (b) that X is fast enough, or (c) that Y is not already fast enough, or (d) that the superior speed of X is a significant difference between the two. A popular example of this second device would be the advertising claim that Hooferin works twice as fast as aspirin. Another form of irrelevancy, commonly referred to as the "bandwagon" technique, implies that because most people do something or believe something (or don't do it or don't believe it), therefore it must be true (or false, as the case may be). Examples of this latter type of device might include such statements as "Fifty million voters can't be wrong"; "Everyone's switching to Doozey, the people's choice in beers"; "More Bulge-O girdles cover more women than all other girdles combined"; "If God had wanted man to fly He would have given him wings." In reality, of course, sheer proportion does not necessarily prove anything; a whole population can be factually wrong about a variety of things at any given time. In science this type of error takes the form of false proof by consensus. Here is one of the knottiest problems of verification, and therefore will be treated fully in Chapter X.

IV.35   (5) *False authority:* The error in this case is a consequence of the factor of halo prestige mentioned earlier. As commonly employed in advertising or political campaigns, this device apparently is quite deceptive; and when employed in science, it sometimes takes the auxiliary form of an overly confident manner of expression or the needless employment of technical jargon merely for the purpose of impressing the listener or reader. In either form, the error is often difficult to establish, for the question of what is a proper limitation or province of a particular authority may not be objectively determinable. Furthermore, the concept "overly confident" is a highly subjective one, just as is the demarcation between a legitimate technical vocabulary and jargon. Nevertheless, the employment of this device should be suspected, especially when it appears in its more blatant forms.

IV.36   The popular exploitation of halo prestige is seen every day. It most often takes the form of endorsements or testimonials (of products, of candidates, or of public issues) by public figures such as movie stars, famous athletes, actors, business tycoons, or any popular person. The fact that the person quoted generally has no more than a layman's knowledge of the thing, person or issue that he is endorsing seems to be unrelated to the fact that his endorsement may prove quite effective in persuading laymen to accept his (false) authority. The employment of a confident manner, and especially of a technical jargon (in reality most often a pseudo-technical jargon) is well demonstrated in the advertising claims of products which range from ordinary piston rings to worthless or even harmful patent medicines ("snake oils"), cosmetics, health appliances, etc.

IV.37 In science, this error generally appears unintentionally rather than purposefully. It sometimes takes the form of questionable authority, for absolute or final authority does not exist in scientific thinking. But it most often appears in the form of sheer jargon—i.e., in the predilection of some scientists to label vaguely recognized but poorly understood phenomena with a large word and thereby imply that the label clearly explains the phenomena. Examples of this tendency might include the following: (a) "Metal fatigue"—implying animistically that metal becomes "tired," though this certainly would be a distortion of the connotation of the concept of fatigue; (b) "Colds"—whatever they may be, as if they are something which possess common attributes; (c) "Group dynamics"—referring to something which seems to occur within and among certain (but not clearly determined or specified) groups of people; or (d) "Sexual inversion"—implying that a person's so-called normal sexual tendencies turn inward (from what?), or develop abnormally—even though there does not yet exist a clear and biologically defensible definition of normal human sexual tendencies. Unfortunately the sheer dynamic quality of science prohibits the complete elimination of this type of error, but careful attention to the principles discussed earlier in this chapter would deter its more flagrant commitment.

IV.38 (6) *Deceit:* Outright lying or deceit is not truly a form of incorrect reasoning, and therefore does not actually belong in a listing of this kind. But it is included here because it is sometimes merged with quasi-deceit, or errors of *exaggeration,* either purposeful or unintentional. No reputable or conscientious scientist ever knowingly employs deceit; but he might easily be tempted, because of his bias or enthusiasm, to exaggerate the significance of his data or of his conclusions. Again, however, the determination of exaggeration is highly subjective, and even competent scientists often disagree on this problem of interpretation.

IV.39 To distinguish a lie one must, of course, know the truth. The layman's vulnerability to this type of conceptual error —essentially, to assume that because a statement is printed or spoken that therefore it must be true—is notorious. The continued efforts of publishers' and broadcasters' associations, Better Business Bureaus, the Federal Trade Commission, the Federal Communications Commission, the Pure Food and Drug Administration, the various libel and slander laws of the land—all of these are of relatively little avail in stopping the torrents of outright lies and blatant exaggerations employed to promote products, persons or issues. Examples that come quickly to mind might include the propaganda of dentifrices that claim to stop tooth decay, of medicines that claim to prevent the wrinkles of aging, of cigarettes that claim to stop coughing, and of various potions that claim to restore "femininity" or "lost manhood." During an election campaign, of course, only the most naive person would believe most of the claims made for or against various candidates; and particularly during times of war, every government tries to convince its citizens that their cause is virtuous while that of the enemy is entirely villainous. In science, true deceit is practically impossible to achieve due to the factor of objective verifiability through replication, although occasionally it may remain undetected for some time (e.g., the so-called Lysenko genetics which became the established party-line in Russia for several years, and the scientistic race "biology" perpetrated in textbooks during the Nazi regime in Germany).

IV.40 Quasi-deceit or exaggeration, however, is something quite different. Here the problem is essentially one of evaluating the significance of a study in terms of the kinds and amount of data and their pertinency, the methodology employed, the conclusions derived, etc. Reference to almost any issue of a scientific journal will indicate that scientists generally are so sensitive about committing this type of error that they tend to err, if at all, on the side of conservatism or caution. Yet the practice of exaggeration among scientists is not unknown, particularly when they are subjected to pressures from practitioners (administrators, technicians, physicians, engineers, etc.) faced with an immediate and pressing problem for which there is as yet no clear answer. In many such cases, programs are initiated based upon small samples, incomplete data, inconclusive (however suggestive) results, or simply in the hope that the hypothesis will be confirmed in practice. Common examples of exaggerated claims occurring in science in recent years might include the claim that atomic energy will soon displace all other forms of power, that death from aging can be forestalled almost indefinitely, that psychoanalytic therapy will cure most mental ills, or that broken homes will be prevented by educating for family living.

IV.41   (7) *False question:* Simply stated, this too-common error generally takes the form either of asking two or more questions as though they were one—thereby evoking either confusion or distortion regardless of the answer given—or of asking leading ("loaded") questions. The common example "Have you stopped beating your wife?" well illustrates this type of error. Basically, this device infers or predetermines the answer to one of the questions, or forces the respondent to give a false impression by his answer. In its more complex form, this device forces the respondent to answer incorrectly if the answers to the different parts of the question are or should be different one from another. Thus to ask "Is gold heavy, yellow, nontarnishable and inexpensive?" is to imply that one answer can (and perhaps should) fit all four facets of the question, when in reality each facet bears no necessary relationship to any of the others. In everyday life, examples of false questions might include the following: "Do you like wine more than beer?" "No." ("Because I don't like either.") "Haven't you dropped out of the Communist Party yet?" "No." ("Because I was never a member.") "Don't you believe in justice?" "Yes." ("But I don't believe that lynching is a form of justice.") In science, examples might include the following: "Doesn't ice float?" (It does in water but not in gasoline.) "Don't stricter laws cut down the divorce rate?" (They may, but it is not demonstrable that happier marriages are thereby induced.)

IV.42   (8) *Incorrect deduction:* A basic type of reasoning rests upon some form of the axiom that things equal to the same thing are equal to each other. (This axiom is examined later in this chapter.) In its distorted versions, however, this axiom can assume several different incorrect forms; for example:

IV.43   (a) *overgeneralization:* If some of A are B, it is commonly assumed that all of A are B. Thus, for example, the demonstrable fact that many women are poor drivers leads many persons (usually men) to deduce that therefore all women are poor drivers. Experience alone, of course, demonstrates the falsity of this erroneous inference.

IV.44   (b) *oversimplification:* If A is part of B, it is commonly assumed that A is all of B. Thus, for example, to argue that the H-bomb is simply an extension of artillery (just as the airplane is contended to be), is to deny the fact that the concept "artillery" is here employed in an extremely narrow and oversimplified definition according to its traditional use. The H-bomb and the airplane are in one sense extensions of artillery, of course, but they are also much more at the same time; and to use the term artillery in this general sense is certainly to distort the significance of its present connotation.

IV.45   (c) *dichotomization:* This type of error results from the tendency to reduce infinite qualities to finite and discrete categories, especially to two mutually exclusive categories. The common example of this tendency is the so-called "black-and-white fallacy," or the practice of asserting that if A is not B, then A cannot be even a part of B. Thus to argue that something is either black or white is to deny the fact that it may be both black and white, or even gray. Common examples of this technique in everyday life might be: "Is he healthy or ill? Is he mature or immature? Is he rich or poor? Is he radical or conservative?" Well-known examples of this type of error in the history of science are the solid-gaseous dichotomy in chemistry, the organic-inorganic dichotomy in biology, the body-mind dichotomy in psychology, the individual-society dichotomy in sociology, the civilized-primitive dichotomy in anthropology, the capitalist-socialist dichotomy in economics, the democratic-authoritarian dichotomy in political science, etc.

IV.46   In science, errors of inference will continue to appear as long as human reasoning is fallible. The three related forms mentioned above all exhibit the basic error of omitting the all-important qualifications "all" or "some," and therefore distort the facts by forcing them into false categories. (This is the same device employed by some lawyers or judges who insist that a witness "Answer either yes or no!"). This type of error can be avoided by the judicious use of qualifications (a point to be treated in the following chapter), and by adherence to the criteria for sound concepts discussed earlier.

IV.47   *(9) Preference for the mean:* The underlying reasoning in this type of error is the belief that the mean or midpoint somehow is preferable to either extreme. This tendency in reasoning possibly results from the exercise of caution, but it might just as easily result from a notion that the midway point is inherently superior to either end. Whatever the rationale, there is no necessary reason why the mean should be inherently more correct than either extreme; and although cautious travel along a "Middle-of-the-road" philosophy may be comforting to timid minds, it can be poor scientific inference.

IV.48   Examples of this type of error as found in popular thinking are quite numerous. To mention just a few, some slogans might be cited: "Not too strong, not too mild, but just right." . . . "All work and no play makes Jack a dull boy." . . . "The middle-of-the-road party, neither radical nor conservative." . . . "The perfect compromise between safe filtering action and real tobacco flavor." As employed in science, preference for the mean can best be noted in much contemporary thinking in the so-called behavioral sciences. Examples might include the belief in educational psychology that a proper (?) admixture of group cooperation and individual initiative results in the most efficient learning situation. In economics the belief is popular that the best form of government is that which exhibits the proper (?) admixture of governmental services and private enterprise. In personality psychology the belief is current that neither biological heredity nor environmental influences plays a distinctive role in shaping human behavior, both are equally (?) significant. In sociology the current belief holds that antisocial behavior is never induced by either individual or cultural factors alone, both are equally (?) influential. In all these and other such cases, preference for the mean might actually deter consideration—and therefore additional research and its resultant knowledge—of the possibility that one of the two factors in question might be much more significant than the other. Therefore to say that both are important (though probably true) begs the basic question: viz., which one is more significant than the other, to what degree, and under what conditions? In science as in daily life, intellectual conservatism should not be confused with timidity; and while avoiding logically indefensible overgeneralizations, it should not be assumed that preference for the mean will insure valid inference.

IV.49   (*10*) *Imperfect analogy:* The most insidious and hence common form of erroneous reasoning probably is the invalid analogy. Based upon the general axiom that things equal to the same thing are equal to each other, this type of deductive inference can take several incorrect forms, only a few of which will be illustrated here. In the schoolboy's version, the argument assumes the form of: "Dogs are animals and cats are animals, therefore cats are dogs." The key problem here, of course, is the meaning ascribed to the term "are."

IV.50   Stating this argument (i.e., logical contention) in its characteristic form, we can call the statement about dogs the "major premise" and the one about cats the "minor premise." These two premises are critical to any conclusions that might be drawn about their relationship. In the first place, these premises do not state whether some or all dogs or cats are being referred to; and in the second place, they do not state whether dogs or cats are only animals or if animals are only cats and dogs. Linking these two aspects of the argument with the problem of interpreting the verb "are", we can illustrate the various types of conclusions that could be drawn from these two contentions as indicated in the diagrams below.

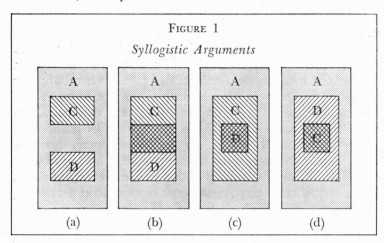

FIGURE 1

Syllogistic Arguments

(a)   (b)   (c)   (d)

IV.51   These diagrams are alike in showing that C (cats) and D (dogs) are totally enclosed in A (animals); and this, of course, is all that is stated in the original premises. The different arrangements of the enclosed squares show the various conclusions that can be derived about the relationships between cats and dogs which are consistent with the premises. The relationships indicated by (a), in which no cats are dogs and no dogs are cats, happens to be true in fact. In (b), however, the relationships in which come cats are dogs while other cats are not dogs, and some dogs are cats while other dogs are not cats—although not true in fact, are equally consistent with the statements in the original premises. The same is true of the relationships shown in (c) and (d). The conclusion asserted in the fallacious inference of the original premises is illustrated by (c): that all dogs are cats.

IV.52   This assertion happens to be untrue, of course; but if the terms designated by the "variables" A, C and D were changed, then the conclusion could be true, for example: All dogs are animals and all collies are animals, therefore collies are dogs. Although this conclusion happens to be true as a matter of fact, the reasoning supporting it is fallacious, because other inferences —conclusions that would be valid—could just as well be derived from the same premises. This is a common example of the tendency to support a true conclusion by an unsound argument—

something that happens often in daily life and occasionally even in the less advanced sciences.

IV.53   The error of invalid inference just illustrated is termed the *fallacy of the undistributed middle* (of a "syllogism"), or in some cases the "all-or-none" fallacy. This fallacy is essentially that of employing terms (in an argument) which are so broad that they admit of various interpretations, or that they permit essentially disparate units (e.g., dogs and cats) to be classified together (e.g., animals). Errors of syllogistic reasoning, therefore, often include the basic error of overgeneralization (as discussed earlier); so it is not surprising to find that overgeneralization is one of the most common, most pernicious and most devastating features of poor reasoning—in or out of science.

IV.54   The alert thinker demands clear and unequivocal definitions, and guards constantly against false analogies or overgeneralized syllogistic arguments. Without practical skill in these various forms of deductive reasoning, however, most laymen and even too many scientists often lead themselves, or are easily led, into logical quagmires which prohibit clear and valid formulations, or meaningful patterns of reasoning.

IV.55   This chapter necessarily has been a long one simply because the improper use of concepts and constructs has bedeviled scientific thinking ever since its beginnings. In glancing over contemporary scientific literature, many critics would probably agree that erroneous thinking—in the form either of poor definitions or poor reasoning—constitutes the prime weakness found among the scientific fraternity. This weakness almost invariably overshadows other types of weaknesses (e.g., of instruments, of techniques, of design, of effort, or of resources) which can impede scientific progress. From the vantage point of hindsight it is often easy (and just as often surprising) to see how the factors which delayed scientific discovery were a consequence of faulty reasoning. Therefore, it is imperative that continued training in, practice with, and checking of one's reasoning processes should be regarded as the indispensable tools of serious scientific endeavor—tools which are easily dulled but which can never be sharpened too keenly. In the chapters to follow some of these same conceptual problems will be referred to again within a new context.

Part II

THE RESEARCH PROCESS

CHAPTER V

Principles of Research

A. FUNCTION OF RESEARCH DESIGN

B. THE ROLE OF THE HYPOTHESIS

C. INTERRELATION OF THEORY AND FACT

A. *Function of Research Design*

V.1 Unlike the function of problem formulation—which is to ask questions that can be answered scientifically—the function of a research design is to organize the procedures of study so that error is minimized, effort is economized, and relevant evidence is gathered efficiently. The way the problem is organized will depend somewhat upon the purpose of the inquiry—i.e., whether it is exploratory, descriptive or analytical. In actual practice these three types of studies are not mutually exclusive, and in many cases a study may combine two or all three of these functions. However, it may be useful to clarify at this point the essentials of these three objectives.

V.2 The main purpose of an *exploratory* study is an examination of a given field in order to ascertain the most fruitful avenues of research. The study may, for example, simply attempt to ascertain the kind (variety) and number (quantity) of elements present in the field of inquiry. It may, on the other hand, seek tentative answers to general questions in order to suggest fruitful hypotheses for research. Or it may investigate the practicability of various techniques to be employed in a given set of study circumstances. In any event, its main emphasis is upon discovery: of problems, of subjects, of techniques, or of areas for more intensive study; and its major attributes are adaptability and flexibility—i.e., it is designed purposely to permit examination of various alternative views of the phenomena under consideration.

V.3 A perusal of contemporary scientific journals would indicate that well-organized exploratory studies are just as meaningful to a growing science as are studies conceived around well-established and clearly formulated problems. Many studies flounder because their foundations have not been adequately prepared through effective exploration. Especially in fields where fruitful theories still remain to be formulated, where many of the influential variables are still to be discovered, where the range and intensity of the units of study are still to be ascertained, and where the magnitude of the investigation is still an unknown quantity—efficient exploratory studies are imperative. Even in those instances where the above-mentioned aspects are satisfied, the function of a good exploratory study is to answer the practical questions of research procedure: what instruments seem to be most effective, what tools will be needed, how much and what kinds of manpower will be called for, how much time will be involved, how much money will need to be spent, or how complete will be the results? In fact, in those relatively unchartered areas of research (which range from interplanetary travel to psychosomatic medicine) where very little is known, many exploratory studies need to be done before a major research design can be attempted.

V.4 A good exploratory study should not, however, be designed as a sort of catch-all of research problems. The principle of parsimony should always be kept in mind. It is much more fruitful to limit the initial exploration to the major problems of research. Depending upon the field or upon the investigator and his purposes, an effective exploratory study might concentrate simply upon one aspect of the total problem. This sole aspect might be the preliminary testing of various alternative hypotheses, of the usefulness of various operational definitions, of specific instruments, of various samples, of subject-matter responsiveness, of replicability, or of skills of the investigators. Or it might simply search for unexpected defects or deficiencies present in any part of the total design through a trial run of the final plan. Whatever the purpose, the good exploratory study is the researcher's best insurance against the ever-present hazard that a large-scale research project might be either vitiated or nullified by the unexpected appearance of unforeseen obstacles. Prudence suggests that a major research effort should not be attempted until the evidence of one or more exploratory studies clearly indicates the feasibility and probable fruitfulness of the final research design.

V.5 A *descriptive* study has as its main purpose the accurate and systematic portrayal of *what is*. It may be concerned with units (people, plants, elements, forces, etc.), or quantities (many or few, one or all, present or absent, etc.), or time sequences (now or later, all the time, etc.), or any other particular features of phenomena being evidentially ascertained and verified. The descriptive study basically tries to answer the questions of who,

what, where, when, or how much; and its essential function is
largely reportorial.

V.6   It might be worthwhile to mention at this point that the
distinction between a descriptive and an analytic study (the latter
to be discussed below) involves a complex semantic problem.
Actually, there is only a very fine line, if any at all, between describing something and analyzing it. For example, to say that
rain occurs when the level of precipitation reaches a certain degree of concentration is both to describe an occurrence and to
analyze a relationship. The same would be true in the case of
describing or analyzing any sequential relationship: the travel of
a rocket, the spread of a disease, the occurrence of a headache, or
the rate of traffic accidents. For all practical purposes, then, the
term "descriptive study" is generally employed when the major
purpose of an investigation is to portray the character, the frequency, or the extent of given phenomena; while the term
"analytic study" is generally employed to denote an examination
of the relationships existing among already-described phenomena. Further details of this rather involved semantic problem
will be considered in Chapter VIII; but it should be borne in
mind that this distinction is purely artificial, and is made simply
for purposes of convenient exposition.

V.7   An *analytic* study is basically concerned with the problem of ascertaining causality—i.e., of answering the question of
"How" or "Why?" (It may also attempt to answer the derivative
questions of "What can be done about it?" or "How can the effect be altered?") It should be remembered, however, that the
question of "Why?" in science is answerable only in terms of
probability relationships, not in terms of ultimate causes. Thus
the question: Why does the light bulb glow when the wall switch
is snapped on? might evoke the response: Because a circuit was
completed, thus causing electricity to pass through the filament
in the lamp, thus causing it to glow. To the further question:
But why does the filament glow when energized by electricity?
the answer might be; Because some materials (in this case, tung-
sten) are relatively poor conductors of electricity; this quality of
resistance causes them to transform the electrical energy into

heat, which in this case takes the form of light. To the further question: But why are some materials good electrical conductors (i.e., have a lower resistance) while others are poor conductors (i.e., have a high resistance)?, the answer might be: Because they differ in respect to their free electrons. In a good conductor, the ring or planetary electrons move freely from one atom to the next; in a poor conductor, they do not—in short, there are few free electrons present, and the substance is thereby a poor conductor. Further explanation would require a description of the electron theory of matter, and hence of the universe as the physical scientist views it in terms of force and matter. The function of the analytic study, then, is to ascertain *what* happens and perhaps *how* or *why* it happens) *when* two or more factors result in a given effect. In other words, its function is to ascertain the meaningful and predictable relationships existing between two or more factors and the variables associated with them. Since there is such an intimate relationship between the notion of causation and the analytic method of study, a detailed treatment of this relationship will be deferred until Chapters VIII and IX—which are devoted exclusively to an examination of analysis and analytic methods.

V.8   Practically every current textbook on scientific method, but particularly those in the social sciences, discusses a wider variety of research methods than the three just mentioned. Among the more common designations are the experimental method, the interview method, the case history method, the statistical method, the questionnaire method, the survey method, the historical method, and the comparative method. Two points should be made here to avoid possible confusion: (a) The terms study and method are not synonymous, though they are often employed interchangeably. A study has as its purpose either a description or an analysis of phenomena, or both; but it may employ any number of techniques to achieve its ends: statistics, questionnaires, cases histories, interviews, etc. (b) The use of any particular technique does not of itself determine whether the data derived therefrom will be either descriptive or analytic, or both—only the employment of those data can determine their character. Of the various techniques mentioned above, two of them (the experimental and the comparative) are inherently analytical, whereas the others may be employed in either descriptive or analytic studies. For these reasons, this book eschews the customary practice of referring to techniques as though they were equivalent to types of studies, and prefers to treat techniques for what they are: tools to be employed in the quest for data.

V.9 Regardless of the type of study to be designed, all research studies proceed through several well-defined steps. These steps are: (a) the formulation of an hypothesis, (b) the explanation of the procedures to be employed, (c) the accumulation of the data, (d) the analysis of the data, and (e) the verification of the findings. Though studies may differ in scope or purpose, these five steps are basic to a well-organized research design. (It should not be assumed that this statement contradicts an earlier statement which states that there are at least eight major steps in the operation of scientific method; for at this point we are concerned only with study design, not with the totality of scientific method.) The remainder of this chapter will be concerned with the first two of these five basic steps, while the following chapters will consider the last three steps. The justification for this arrangement inheres in the fact that steps one and two (formulation of the hypothesis and the rationale of the procedures) are necessarily interrelated—i.e., one cannot be functionally considered without reference to the other. The third step (viz., accumulation of the data) is largely mechanical, in the broadest sense of the term, in that data may be collected without any necessary reference to their employment in a particular research design. This third step, however, requires three chapters for adequate explanation and illustration. Step four (analysis of the data) ties the data to the first two steps (i.e., to the hypothesis and to the procedural rationale); and in conjunction with the fifth and final step (verification), requires a whole chapter in its own right. Ignoring at this time the incidental but functionally unrelated problem of data presentation, the following discussion will be concerned with a consideration of steps one and two.

B. *The Role of the Hypothesis*

V.10   For the time being a *theory* may be defined as a generalized statement or proposition which explains or interrelates a set of other more specific propositions. Some facets of the total interrelationship may be clearly perceived, others may not; some sub-relations within the total relationship may be definitely accounted for, others may not. In order to utilize a theory, therefore, inferences are deduced from it. Such inferences constitute hypotheses. An hypothesis, then, is a proposition derived from a theory by deductive inference, and which permits a test of empirical confirmation. Its function is to extend verified knowledge beyond the present borders of any given field of theoretical knowledge. As such, the hypothesis is not only the link between speculation and verification—but more importantly, it is the essential "growth factor" of all scientific knowledge.

V.11   It may be helpful to clarify the function of hypotheses by classifying them into two basic types in terms of their level of abstraction. First, some hypotheses state the existence of empirical uniformities—i.e., they state facts or relationships which may be commonly known (so-called "common sense notions"), but which have not yet been verified empirically, or which are not yet known but are suspected to exist. These are often referred to as descriptive hypotheses. In this group would be included problems of degree of uniformities—for example, one might ask in hypothetical form: "Does lead weigh more than tin?" "Do more men than women catch colds?" "Does this type of bug respond to this type of spray?" "How many Republicans are also Protestants?" Although simple descriptive studies may be pursued without reference to a hypothetical framework, much data gathering implies an underlying hypothesis. Thus, for example, to say that one is simply interested in finding out how many people drink alcoholic beverages is to ignore the fact that the inquirer wants to know for some particular reason; and that that reason probably is related to a general theory from which such hypotheses as the following might be deduced: (a) people who drink are of certain types in terms of age, sex, race, occupations, etc; (b) they drink as much as they do for particular reasons, or (c) under certain conditions, or (d) with certain patterns of consistency, etc.

V.12 Since a basic attitude of scientists is skepticism, it might be well to mention at this point that many so-called common sense assumptions are either false in fact, or seem common sensical to the layman only after having been empirically verified. A good deal of popular scoffing at scientists stems from the layman's inability to understand the significance of testing even the simplest inferences about relationships. Yet many popular hypotheses turn out to be unfruitful because they assume truths which have not been empirically verified. Examples of such unverified assumptions might be the following: (a) that increasing formal education leads to increasing "intelligence" (defined as the ability to make more rational choices); (b) that "you get what you pay for" (i.e., that quality is always directly related to price); or (c) that "tough times produce tough leaders" (i.e., that effective leadership arises when the situation demands it). Examples of invalid common sense assumptions might be the following: (1) that "human nature is pretty much the same all over the world;" (2) that personality is genetically transmitted ("like father, like son"); or (3) that erotic potential is inherently greater in males than in females. Since science attempts to achieve a systematic body of verified knowledge, a large part of the total endeavor is spent in verifying assumptions which may seem obviously true—especially to those having the hindsight of "Monday-morning quarterbacks"—but which have not been empirically verified.

V.13 The second type of hypothesis is concerned with the relationship between variables. This type is often referred to as an "analytic hypothesis." It seeks to discover the degree to which a change in one factor is related to a change in another. The number of variables correlated with the observed uniformities in the data, or with each other, depends upon the purposes of the study; but in general it is far easier and safer to test the influence of only one variable at a time. It is far easier, for example, to try to understand the influence of rapid temperature changes

upon the contraction of colds than it would be to try to understand the influence of differing temperature, humidity, diet and
age upon the probability of contracting a cold. Especially in an
area where relatively little is known, caution suggests that small
bites taken into the pie of knowledge are less apt to result in
mental indigestion.

V.14 Many of the attributes of analytic hypotheses are similar
to those of constructs (discussed earlier), for the simple reason
that hypotheses are composed of concepts or constructs. Essentially a useful hypothesis is one (a) which is conceptually clear
and unambiguous; (b) whose terms are related to objective and
empirical referents or operational definitions; (c) which is specific and simple (the latter quality in order to promote parsimony); (d) testable (i.e., related to available techniques); (e)
plausible (i.e., it should meet the test of logical possibility); and
(f) meaningful.

V.15 Qualities (a), (b) and (c) above have been dealt with in
the previous chapter. Quality (d), testability, refers to that attribute of a hypothesis which makes it amenable to available techniques. If a hypothesis asks a question—or, more accurately, states
a factual empirical proposition—which cannot be tested, then
research cannot proceed. Examples of hypotheses which are non-
testable because techniques are not now available might include
the following: (1) the pain of a toothache is potentially greater
than is that of an earache; (2) mankind will recover after an "all-
out" atomic war; (3) lowering the voting age in all states to
eighteen years will result in better political choices.

V.16 Quality (e), plausibility, refers to a test of logical possibility. This is, at times at least, a highly subjective quality.
What might seem plausible or logically possible to one person
might seem just the reverse to another. Successful invention, for
example, is often the result of entertaining what at the time
would appear to most persons to be an implausible relationship.
What so-called normal person would have suspected, for example, that a powerful force could exist without having any weight,
shape, size or volume? (viz., electricity); that a human voice
could be transmitted through space without any visible or physi-

cal medium of travel? (viz., radio); or that man might travel through outer space? Quality (f), meaningfulness, is possessed by an hypothesis which, if confirmed, is productive of significant implications as stimuli to further research; especially when it has a unifying function—i.e., when it is fruitful in enabling the scientist to bring more facts into a meaningful system or order of explanation. When a theory suggests various and sometimes even apparently rival hypotheses, the criteria listed above are employed to suggest which particular hypotheses should be pursued.

V.17 The test of an hypothesis follows the pattern of analytic methods to be discussed in Chapter IX. During the formulation stage, however, an hypothesis must include within itself its criteria of confirmation. That is, it must be stated in such terms that there would be no serious question about the answer offered as confirmation. This means, in practice, that a good hypothesis is generally stated quantitatively; e.g., "This chemical will lower the temperature of distilled water at least two degrees Centigrade for every gram in every liter." The criteria inherent in the statement of an analytic hypothesis define both the type and degree of confirmation desired in order to pass the test of acceptance.

V.18 It should be pointed out, however, that a hypothesis is never definitely or absolutely proved; for if it were, knowledge could not advance beyond that point. Rather, a hypothesis is *confirmed* or not confirmed. In this sense the type of confirmation must be clearly stated in the results of the test. The idea of degree of confirmation suggests that the verification of an hypothesis is expressed in relative, never in absolute terms, and thus is to be stated in terms of some degree of magnitude according to some standardized method of designation. When all the functional and implicational features of a hypothesis are visualized in the context of the total scientific effort, it is not surprising that the formulation of a good hypothesis is regarded as the *sine qua non* of fruitful research.

V.19 A very practical problem which often arises to vex the scientist is the question of how long to pursue a hypothesis which seems to defy confirmation. Obviously there can be no definitive or categorical answer to this question. In the first place, there is a great deal of difference between stubborn adherence to an idea which is not tenable in the face of contrary evidence, and perseverance with an hypothesis which is very difficult to demonstrate but against which there is no direct evidence. (The difficulty, of course, is to determine what constitutes contrary or indirect evidence.) Two simple examples would be the case of the unfruitful but fundamental hypothesis of the alchemists (viz., that base metals could be transmuted into gold); and the case of the presumably even more implausible hypothesis that man could learn to "fly."

V.20   In the second place, the validity and pertinency of the facts themselves are often open to legitimate question; and the best minds often disagree on facts as well as on their relation to an hypothesis. (For example: Is it a fact that young women tend to develop more of their potential abilities in sex-segregated schools than they do in coeducational schools?) In the third place, when is the point reached at which the promising hypotheses are irrevocably destroyed by the evidence of cold-blooded facts? Here is an area of wide differences of opinion even among famous scientists. Perseverance has resulted in many spectacular discoveries, but so has the willingness to relinquish cherished hypotheses in favor of more promising ones. Personal conviction, emotional, intellectual or financial involvement, need for results, time and facilities available, degree of psychological open-mindedness—these are only some of the factors which operate to influence the resolution of this problem.

C. Interrelation of Theory and Fact

V.21   Laymen are prone to think in dichotomous terms (viz., the "black-and-white" fallacy discussed in Chapter IV). It is a common practice to set up rigid distinctions between theory and fact or between theory and action. Actually, of course, theory, fact and action are simply different but interrelated aspects of the same general behavior pattern. It might be worthwhile at this point, therefore, to clarify the role of fact and theory as these concepts are employed in science.

V.22 A scientific *fact* is an objective or empirically verified observation or deduction. As such, the fact is the basic element of reliable knowledge. But in science facts are not employed randomly; rather, they are interrelated in apparently meaningful ways to suggest causal relationships. These relationships constitute "theory." Therefore, facts may be employed (a) to suggest theories or new relationships ("discoveries"); (b) to suggest revision or rejection of existing theories; or (c) to redefine or clarify theory. It should be clearly noted at this point, then, that irrelevant facts cannot be arranged into valid theoretical relationships; and therefore that unverifiable theories most often exhibit a foundation of unverified facts.

V.23 The relation between facts and theories is not always direct. An interrelated set of verified facts may be formulated into an empirical regularity usually defined as a "law." A *theory*, however, is a generalized, synthetic explanatory statement of the "cause" of a phenomenon or of the interrelation between classes of phenomena. As such it often employs abstractions having no apparent empirical qualities (e.g., "force," "symbiosis," "intelligence," "social mobility"). Its function is to serve as the unifying explanation for an unlimited series of possible deducible hypotheses; just as it may "explain"—or systematically account for—the relationships among laws.

V.24 In this sense, then, facts do not necessarily prove or disprove a theory any more than a theory "justifies" certain facts. The intermediary between facts and theories is either the various hypotheses deduced from a theory, or the experimental laws tying together verified facts. In some cases a theory may be so abstract as to defy deduction of fruitful (i.e., empirically testable) hypotheses. (Many theories in the social sciences, for example, exhibit such levels of abstraction or even abstruseness—e.g., marginal utility, social evolution, socialist efficiency, unconscious motivation.) In other cases a theory may be only one small conceptual step removed from an empirical law (e.g., the theory that value is related to scarcity). In any case the development of fruitful theories encompassing the widest possible range of phenomena (so-called "grand" or "unified" theories) is the hallmark of an advanced science.

V.25   If facts may be regarded as the foundation of reliable knowledge, theory may be regarded as the superstructure. As employed in science, theory (a) serves to orient study: it narrows the range of facts to be utilized and at the same time determines which kinds of facts will be deemed relevant to the purposes of the study. (b) Theory also serves as a system both of conceptualization and of classification. It permits the creation of concepts which refer to major processes (e.g., "forces" or "dynamics"), and the classifying of relevant objects (taxonomy); and it permits the creation of a structure of concepts ("conceptual schemata"). (c) Theory permits a summarization of what is already known about a phenomenon, thereby further permitting (1) a statement of empirical generalization, or (2) the creation of systems of relationships between propositions ("laws," "principles," "axioms"). (d) Theory may suggest the prediction of facts; and therefore (e) point out gaps in existing knowledge.

V.26   The intimate and significant interrelation between facts and theories can be illustrated by reference to just one of the functions of theory mentioned, to wit: conceptualization and classification (see b, above). Now, the same facts obviously can be classified according to a variety of categories. A man, for example, can be classified in terms of his physical features, his political preferences, his religious affiliation, his academic achievement, his occupation, etc. But there is no logically necessary reason why one type of classification is inherently superior to another; it depends upon the function which the classification is to serve. Yet the function of a classificatory system can be rationalized only within the framework of a given theory of behavior—i.e., it is the theory which both justifies and gives meaning to the particular classification of facts.

V.27   Thus, for example, automobiles are classified according to their horsepower, their piston displacement, the number of cylinders, their body types, their gross weight, the length of their wheelbase, their initial cost, their operating efficiency, their used-market value, etc. Each one, or several in combination, of these attributes serves some useful purpose in terms of some function; and that function is related to a theory of meaningful behavior (e.g., a theory of the cost of maintenance, of resale value, of obsolescence, of ease of parking). It would seem difficult, however, to justify the classification of automobiles in terms of such facts as that some are washed more frequently than others, that some have ashtrays in one position rather than another, that some have larger air-filters than others, or that some have shorter names than others. In all of these cases, however, it might be discovered that a particular and previously ignored fact could become the basis of a significant classification—provided that a theory of causal connection could be established that would relate such a fact to another meaningful fact (e.g., that for many women, sales appeal apparently is strongly influenced by the color of the upholstery).

V.28 The above remarks suggest another reason why the common distinction between description and explanation is largely superficial. Description is meaningful only when the phenomena being portrayed are described in terms of significant attributes—i.e., in terms of facts chosen according to a meaningful criterion of classification. The commonplace statement: "Let's get the facts and then see what they suggest," is meaningless if taken at face value; for it does not indicate which kinds of facts should be secured. In short, such a statement does not indicate the ways in which the attributes of a phenomenon are to be perceived, therefore collected, therefore classified, and therefore related.

V.29 The popular confusion regarding the relationship between fact and theory has several significant implications. First, the so-called practical person who scorns theory simply does not realize that most actions are a consequence of theoretical formulations, which formulations serve as the justifications for the actions. It follows, therefore, that anyone who believes in the relation between facts and actions is a theorist (in the general sense of the term), whether he realizes it or not. For example, the person who argues that "actions speak louder than words" is saying, in effect, that actions occur independently of the ideas which motivate them, or that actions (necessarily?) convey specific meanings to their recipients, or that people are always more re-

sponsive to nonverbal behavior. Yet the rationale of any deliber-
ate action is always some supportive idea, whether clearly formulated or specifically recognized or not; and the supportive idea
itself is derived from some theory of responsive behavior. Thus
the "believer in facts" is saying, in effect, that such facts are valid
and meaningful because they rest upon some theory of significance, credibility or function, even though he may not have
reasoned through the surface of the fact to its origin in the
theoretical notion which supports it.

V.30 Second, speculation about facts is impossible without
theorizing—at however simple a level. To speculate about facts
is to manipulate the significance of those facts in relation to other
facts. But the significance of the interrelation among facts implies some kind of *causal connection,* hence some theoretical
notion. Thus the fact of death, for example, cannot be utilized
meaningfully by itself because it has no behavioral significance
by itself. It acquires significance only when it is arranged meaningfully with other factual notions about reality. It may, for
example, be meaningfully linked (a) with a notion of a desirable
condition (e.g., "I wish he'd drop dead"; "I'm glad I'm not
dead;" "I hope this snake is dead"); or (b) with a notion of causal
connection (e.g., "He died because someone shot him"; "He'll die
if we don't stop the bleeding"; "His death will end the family
line"). All such speculations require theoretical constructs about
possible or probable reality; facts do not "arrange" themselves
in the mind by some natural or inevitable process.

V.31 Third, the great majority of decisions which constantly
affect everyone's life in politics, religion, law, occupation, economics, love, etc., are based upon theoretical considerations—
whether those considerations are formulated as such or are simply implied by the consequences of the decision. Except in purely
inborn (inherent), habitual or randomized reactions, decisions
indicate a choice between alternative possibilities. Recall the
conditional situations of "If this, then that," as discussed in
Chapter II: however practical or pragmatic, *a decision is a conclusion drawn from a conditional argument.* As such, it indicates
a sequence of inferred relationships between propositions. The

final choice, however useful or not, cannot be derived from a vacuum of ideas; it can only be derived from the set of assumptions held by the person making the decision.

V.32   Fourth, the so-called realist who believes only in what he himself sees, hears, feels, etc., or the so-called pragmatist who has faith in something simply because "it works," both fail to realize that the very definition of the reality of impressions, on the one hand, and of the functional utility of an act, on the other, are both determined by the theoretical notions which exist in the mind of the person.

V.33   In other words, to say, for example, that "I believe what I see" or "I know that it's good because I proved it to myself by trying it out," is to contend, in effect, (a) a belief in the dependability of one's sense impressions, on the one hand, and (b) a faith in one's tests of verification on the other. But both these notions are not spontaneously generated; they are derived from previous experiences with reality and the resultant interpretations of those experiences. In either case, an examination of such statements would reveal that they rest upon theoretical notions related to other notions about empirical or functional reality and truth. The lack of a verbalized formulation of such ideas or idealizations should not be assumed to imply the lack of their existence in one's mind, however unverbalized such constructs may be.

V.34   Thus the so-called realist who contends that he believes what he sees is arguing, in effect: "I see clearly, accurately and meaningfully; therefore this thing I see is clear, accurate and meaningful. It is meaningful because I know its inherent significance; and its inherent significance is so-and-so." (All of these contentions could, of course, be verified.) The so-called pragmatist, likewise, would be arguing, in effect: "A radio that plays is a good radio, and this radio plays; therefore it is a good radio." But like the so-called realist, the pragmatist also supports such notions ("plays," "good") *only by* derived theories of how much and what kind of playing should be expected of a "good radio"; and such *theories,* however unverbalized or unrecognized as such, underlie all arguments related to reality, function and truth.

V.35   In summary, then, it may be seen that fact and theory are dual aspects of the scientific quest for verified knowledge. Each has its place in scientific endeavor. Some scientists spend the major portion of their efforts in the discovery of facts; others concentrate on the formulation of theories. Each function is indispensable to the other, whether the two functions are performed by the same or by different persons. It is true that unsupported theories serve no useful function in science, just as unverified facts do not. But it should always be borne in mind that what is unverified today may be verified tomorrow, and that some of mankind's most apparently abstract theories (e.g., $E = Mc^2$) later were demonstrated in a most dramatic manner (e.g., the atomic bomb).

D. Sampling

V.36   A good research design is in effect both a blueprint and a complete set of construction specifications. It specifies not only what is to be done but also explains why it is being done as it is (and perhaps why it is not being done some other way). Each procedural step is explained so that any competent student of the subject can check the study from its inception to its conclusions. Units, definitions, techniques or procedures that might be questioned as to their validity or appropriateness are justified according to the author's "frame of reference" (see Chapter IX, D) so that his rationale may be clearly understood. In many instances—depending upon the state of accepted knowledge in the given field—much of this explanation is unnecessary because the rationale underlying the study is commonly accepted by most scholars in the field; but if there might be any doubt about either the theoretical foundation or the procedural superstructure, potential criticism is anticipated and met wherever deemed practicable throughout the study.

V.37   After the study design has been explained, the next step is the determination of the sample to be employed. Since it is a rare study that can ever hope to examine all the units of a given *universe* (i.e., a defined grouping of units for study), sampling is the standard and only feasible method for studying classes of phenomena. In some few instances all the units to be studied may be available and actually utilized. (Oddly enough, in such cases, the total population would still be referred to as "one-hundred percent" sample—even though the term "sample" in such instances clearly is paradoxical.)

V.38 A *valid sample* is a portion of a universe selected for its representativeness and adequacy in terms of the specific attributes being considered for study. Since a good deal of statistical knowledge is required in order to determine the validity of a sample, only the essential features of it will be treated here. Let us consider, for example, a sack filled with pieces of coal all broken from the same block. If it had been established previously that the original block was homogeneous throughout, then any single piece of coal would be a *representative* sample of the entire mass in terms of the quality of that mass. If, to continue the example, the original homogeneous block of coal had been sawed into identical pieces, then any single piece would be an *adequate* sample of all the other pieces. Representativeness, then, denotes the *qualities* of the universe contained in the sample, while adequateness denotes the *distribution* of those qualities within the sample.

V.39 Since a good deal of methodological error in scientific research stems from invalid sampling, it may be appropriate at this point to illustrate an invalid sample in terms of both its representativeness and its adequacy. Let us consider in this case a truckload (the universe) of "dry mix" materials consisting of one part of cement, two parts of sand, and four parts of gravel, and then let us visualize a worker scooping out of the pile a bucketful of the material. Suppose that the three units (cement, sand and gravel) had been poorly mixed. Then a given bucketful (sample) might contain all three units—and therefore would be representative in terms of quality—but not in proportions comparable to the original universe, and therefore would be inadequate as a sample. Suppose further that the truck had been loaded in layers: first gravel, then sand, then cement. A given bucketful now might contain only cement, and therefore would be unrepresentative of the universe because it did not contain the other two units; but it would also be inadequate because it did not contain a comparable proportion of all three units found in the original universe (the truckload).

V.40 The validity of a sample, therefore, can be determined only after a meaningful relationship has been established between (a) the quality and the quantity of the attributes being studied, and (b) the hypothesis being tested. This means, in effect, that in many cases an exploratory study must first be done in order to ascertain which attributes of the universe seem to be significantly related to the phenomena under consideration, and also whether the quantity of those attributes bears any significant relationship to the effects being studied.

V.41 The *size* of a sample, therefore, is often a critical determinant of its validity. Fortunately, modern statistical procedures permit a rather accurate determination of sample validity based upon formulas now well standardized. Such formulas are based on probability theory; and permit a rather specific designation of the size of the sample necessary to be obtained when the universe is given and its essential elements have been delimited. It should be obvious, therefore, why it is impossible to designate in general terms how large any sample should be before it can be assumed to be valid, and why knowledge of statistical theory and techniques is basic to any understanding of sampling. Suffice it to say that in a perfectly homogeneous universe or population, a single unit would constitute both a representative and an adequate sample; but in a completely heterogeneous mass, the only adequate and representative "sample" would be one hundred percent (i.e., all) of the units.

V.42 *Random sampling* refers to the process of selecting from a universe a sample in which all units have an equal opportunity to be selected; but it does not mean that one just chooses the sample haphazardly or "at random" in the popular sense of that term. In practice any number of methods may be employed to select a random sample: e.g., every tenth name in the telephone book (if all the names in the book constitute the universe); every fifth shopper (if all the shoppers constitute the universe); every woman who eats lunch at this counter (if these women and this counter are assumed to be representative of the universe—i.e., "women at lunch counters like this"). Or tables of random numbers may be employed which have been statistically devised to assure unaffected scrambling of all ranges of numbers.

V.43 A *stratified sample,* on the other hand, may be preferred in those instances where the distribution of the relevant attributes of the universe is known. In such instances a sample is chosen which is known to possess the qualities of relevance, and proportionately to the distribution of those qualities in the universe. Suppose, for example, that it is desired to make a prediction of voting behavior. According to the theory (let us assume), the relevant factors which influence a person's voting choice are: age, geographic region, religion, occupation and education. The sample in this case would need to select a representative proportion of the total voting population in terms of these specific attributes. If "professionals," for example, should constitute five per cent of the total occupational grouping of the voting population, then the sample should contain five per cent drawn from professional ranks. The same procedure would be followed for all the other attributes. The size of the final sample, therefore (in this same illustration), might be no larger than a few thousand persons—provided that they finally exhibit all the relevant qualities proportionate to the entire voting universe.

V.44 The various subsamples are then combined by selecting among them according to some system of random selection or by employing a table of random numbers. Although the major superiority of stratified over simple random sampling is precision (i.e., reliability), in many instances the extra precision obtained is not demonstrably worth the extra effort involved. The reason for this lack of worth lies in the fact that a random sample includes all possible variables, whether suspected or not; while a stratified sample may err by being overly selective.

V.45 Before leaving this discussion of sampling, it might be worth pointing out that statistical procedures are not necessarily linked to sampling alone. Statistics is, after all, a method of expressing quantified relationships—either for purposes of de-

scription or of inference—but those relationships are not necessarily limited to simple and discrete units. In the realm of complex theoretical formulations which are commonly employed
on the frontiers of the physical sciences, statistical formulas are
often employed to express elaborate concepts and constructs
rather than just simple, discrete quantitative units. It is questionable, therefore, whether the employment of such higher mathematical formulations in advanced research properly should be
designated simply as "statistics." True, both methods involve
mathematics, but in different ways and for different purposes.
The higher mathematics (especially algebra, trigonometry and
calculus) employed in the advanced sciences—especially in the
form of "symbolic logic"—bear relatively little relation to the
kind and level of mathematics ordinarily employed in statistical
procedures. This is perhaps another way of saying that science
employs systems of *symbolic logic* which permits the manipulation of the most abstract and intricate concepts devised by man's
intellect.

E. Delimitations of Data

V.46 Somewhere after the initiation of a study design but
soon after its universe has been selected and its sample determined, serious thought must be given to the limitations that the
researcher intends to impose upon himself in terms of his concepts and his units of study. (In other words: Are the concepts
specific enough, and are the units sufficiently precise for the kind
of study intended?) Several references already have been made to
the necessity for defining concepts clearly and unambiguously.
In many instances, however, it is semantically impossible to
designate an idea with a single, precise word. Throughout a
study, therefore, six general classes of *qualifications* are necessary
in order to promote clear and exact communication. The first
class of qualifications designates the degree of *generality* referred to. Words such as some, a few, all, rarely, sometimes,
usually, always, never, generally, the average, etc., are employed
in order to facilitate both understanding and clarity. Wherever

possible, however, such qualifications should be employed to increase precision, not to negate it; and should have such standardized meanings that comprehension rather than confusion is engendered by their use.

V.47 The second class of qualifications relates to the *time factor* involved. Depending upon the relevancy of this factor, it might be necessary, on the one hand, to specify the time in fractions of a second, in minutes, in days, in months, in years or in centuries. On the other hand, it may be desirable to generalize within definable limits—for example: at all times, since the beginning, after the initial effort, during this present series of studies, during the past generation, until the end of this cycle, etc. Whatever the degree of generalization, the general principle of relevancy and exactness should be adhered to as determined by the purposes of the study.

V.48 The third class of qualifications designates the *area* being studied. Does the study as a whole, or just its sample, or just its effects, apply (for example) to the entire solar system, or to all this earth, or to a particular nation, or only in valleys, or within certain geological strata, or in large cities, or only on Third and Main streets? Ecological studies (i.e., studies of the relation of an organism to its environment) by definition are necessarily place-oriented; but very few studies, particularly in the social sciences, are meaningful without a specific designation of the area being encompassed.

V.49 The fourth class of qualifications relates to the units or *population* being studied. In the physical and biological sciences, this type of designation does not generally present a problem for the researcher, since studies in these fields tend usually to be quite specific in the designation of their population. Occasionally, however, a deficiency in specificity can be noted even in these fields. Bacteria, for example, are of many kinds, and so are forces, plants, cells, and other such classes of phenomena having specific attributes. By and large, however, the notorious deficiencies in this fourth type of qualification have occurred in studies conducted in the social sciences. "Man," for example, has been a notoriously ambiguous unit of designation. At times he is implied to represent all humans who have ever lived, or all those now living, or males only, or White Protestant Southern males only, or young married men only, or— in a famous study purportedly of sexual behavior in the human male—only a highly selected and small group of statistically deviant subjects. It is imperative, therefore, that the group referred to must be qualified as rigidly as the hypothesis of the study demands.

V.50   The fifth class of qualifications denotes the *conclusiveness of the evidence.* A scientific study can never conclude with such a statement as: "This study finally proves for all time. . ." In most instances the evidence *suggests* rather than *proves;* and even then only with such accompanying qualifications as: "On the basis of this study. . ." "According to this sample. . ." "According to preliminary studies. . ." "It appears from this evidence. . ." "It seems at the present time. . ." etc. Such statements should not be viewed as condescensions, nor as indicating a false intellectual humility but, rather, as admitted safeguards against the potential dangers of overgeneralization.

V.51   The sixth and final class of qualifications indicates the *extent of professional agreement* relevant to the various methods employed in the study or to the conclusions drawn from it. Since consensus among the fraternity of scientists is often an important indication of validity, each researcher is obligated to denote the extent to which he agrees with his fellow scientists and they with him. Whether he knows them personally or not doesn't matter; "knowing" in this sense is a result of reading the professional literature and of being aware of the developments in the field. Therefore, whenever a question might reasonably be raised about the degree of professional agreement, it is a matter of both intellectual honesty and avoidance of unnecessary criticism to qualify statements appropriately by such remarks as: "According to all/most/some/a few/ the majority of scholars in this field. . ." or "Without exception. . ." or "With few exceptions, particularly among . . ." or "Except among adherents to the . . ." or "According to the proponents of . . ." etc. Like other types of qualifications, this one protects the researcher from possible charges of generalizing beyond the limits of his frame of reference or of his data or method.

V.52 This chapter has been concerned mainly with the theoretical core of scientific study and research. The following two chapters will examine the basic problems and principles relating to the collection and evaluation of data. The basic principles underlying scientific research generally evoke the greatest amount of interest in the scientific literature; for particularly with respect to the role of theories and hypotheses, the success or failure of any scientific study depends most directly upon the theoretical foundations upon which it is built. It is for these reasons that the major interests of "pure" research or "pure" science revolve around theoretical formulations and constructions—even though the spectacular evidences of scientific knowledge are seen only in its applied aspects. Without a well developed body of theory, however, no branch of science ever proceeds beyond a primitive level of achievement.

CHAPTER VI

Data Collection: Preliminary Considerations

A. ROLE OF THE LITERATURE

1. *Purpose of literature review*
2. *Seeking the best avenues of approach*
3. *The use of models*
4. *Suggested tools, techniques or instruments*
5. *Avoiding pitfalls or blind alleys*
6. *Achieving insight to possible solution*
7. *The literature as data*
8. *Role of personal correspondence*
9. *Locating relevant materials*
10. *Subjective aspects of relevancy*

B. EVALUATION OF MATERIALS

11. *Relevancy related to validity*
12. *Problems of validation*
13. *Validity and reliability*
14. *Examining historical data*
15. *Internal criticism of historical materials*
16. *Historical interpretation*
17. *Role of a bibliography*

C. INSTRUMENTS AND TOOLS

18. *Choosing effective and appropriate devices*
19. *The limitations of tools and instruments*
20. *Ingenuity and inventiveness*
21. *The function of instruments*

A. Role of the Literature

VI.1 The preliminary phase of data collection is an exacting scrutiny of published materials. How useful the existing literature will be depends upon the degree of originality of the given study to be conducted. Some scientific studies are simply replications of previous studies, done for the purpose of corroboration; others, however, are virtually pioneers in a given area of interest. Particularly in the latter cases, examination of the literature in that field is imperative for several reasons. The first and basic reason is to prevent needless duplication of work already done. It is not too uncommon for an enthusiastic novice in science to believe that his (to him) original idea, method, technique or datum is truly original, and then ruefully to find later that it is common knowledge to all those experienced in his field. Duplication may be justified in its own right, especially for purposes of replication; but unless recognized as such, represents simply wasted time and effort. True originality in any branch of science is rare; but the novice is often naive enough to believe that nothing significant has ever been done in his selected field of interest.

VI.2 The second major reason for examining the literature is to seek the best avenues of approach. The problem of study design, or organization of strategy (as contrasted to data-gathering techniques), is a critical one in all fields of endeavor. Like the commander of an army attempting to achieve an objective, the researcher has to decide which plan or approach (strategy) and body of techniques (tactics) seem to hold the greatest probability of success. This decision is crucial in all scientific research; and both ingenuity and experience play a large part in its determination. The literature, therefore, might indicate the relative fruitfulness of particular approaches, and therefore suggest alternative strategies or tactics. Especially when one has worked long and intensively in a given area, it often becomes increasingly difficult to visualize, let alone invent, a new approach to an old and baffling problem. It is here that knowledge of the literature—even, occasionally, of that of related fields—might prove to be of critical value in suggesting the solution to a problem.

VI.3 The best avenue of approach is oftentimes suggested by a relevant *model* found in the literature. A model, as employed in scientific discussions, refers to a parallel form, simpler than the complex subject it represents, but having a similar structure or organization. Models are used to help our minds understand the arrangements of key components and how they function. In the physical sciences, for example, it is quite common to explain the structural organization of molecules by creating physical models showing the number and arrangement of their constituent atomic elements. In actuality, of course, the hydrogen atoms, let us say, are certainly not green nor are the oxygen atoms red; nor is their arrangement static; nor is their size truly represented by the green and red wooden balls used to construct the model. Likewise, a given experiment in a psychological laboratory may serve as a model for further studies even though it will not be copied or duplicated exactly. Based upon analogy and heir to all the pitfalls of analogical reasoning (as discussed in Chapter IV), the search for a model may nevertheless prove quite fruitful to one seeking the most efficacious form of study design.

VI.4 The third function of the literature is to suggest which specific techniques, tools, apparatus or instruments might be most useful in terms of the data to be sought. Though an experienced researcher would normally be quite familiar with all the common devices employed in his field, a search of the literature might suggest new applications, new refinements, or even new interpretations of existing techniques. This aspect of the literature-check is always necessary, even to an experienced researcher, in order to assure that no new and relevant idea is ever left unexploited.

VI.5 The fourth function of the literature is to assist in avoiding possible pitfalls or blind alleys along the road of successful research. It is not uncommon, for example, for a researcher to believe that he has devised or discovered a highly promising theory, fact, design or technique only to read that the same idea proved impractical or unfruitful in someone else's experience. Due to common interests and type of training, independent invention (of ideas, techniques, etc.) often occurs in science; so it is only prudent as well as efficient to be aware of another's difficulties or failures with a proposed line of research. The novice particularly, and often the layman, can easily be overly impressed with his "brainstorms" because he has not been aware of the impracticality or invalidity of such ideas. Such awareness is achieved only by a thorough knowledge of the literature.

VI.6   The fifth function of the literature is to assist in that most difficult of all aspects of research: viz., achieving insight into possible solutions to a problem. It is axiomatic in science that the most difficult of all problems is that of asking the right questions. But it is just as axiomatic that discovery or invention (of theories, methods, techniques, etc.) is a result of superior or fortunate insight into heretofore baffling relationships. (Insight, as employed here, refers to the ability to recognize or create relationships among phenomena.) In this sense insight is often aided by perusal of the literature. Though the literature may not contain the specific answer sought, it may provoke fruitful speculation eventuating in a critical insight (*Geistesblitz,* or lightning-like idea). Such critical insights are the essential substance of discovery and invention. It is also axiomatic, therefore, that science—especially in the more undeveloped fields—would be farther advanced if all the useful information already existing in the literature should be exploited to its fullest potentialities.

VI.7   The sixth function of the literature is to provide data. Such data are of two major kinds: (a) those which are normally found in reference sources (e.g., population data, economic data, weather data, geographical data), and (b) those specialized kinds relevant to a particular study. Whichever kind is sought, overlooking useful data already on file is an inexcusable deficiency in any scientific study. Skill in the exploitation of library resources is an indispensable quality of any competent scholar, and the scientist is nothing more than a specialized type of scholar. Without the ability or the interest to search for data already existing in the literature, a scientist is bound to fail in even the most modest research design.

VI.8   In fields where knowledge has become very proliferated and research quite voluminous, even an examination of the existing literature may not be sufficient insurance that the researcher is abreast of all current developments. In such cases a further step in background preparation is necessary: viz., personal correspondence with other scientists engaged in the same field of study. This procedure may be necessary also in those cases where the available literature is so meager that for all practical purposes it is fruitless. In either case the competent researcher makes every possible effort to familiarize himself with every bit of existing relevant knowledge either during the initial stages of his study while he is still formulating his hypothesis, or at the time when he begins to gather his data.

VI.9 The problem of locating relevant material already existing is not a simple one. Many great discoveries have resulted from a synthesis, or a new synthesis, of facts already known but unrelated to each other. Examples which come readily to mind are: Archimedes' principle of buoyancy—which linked in a meaningful relationship the already known facts that certain objects float while others sink, independently of their size or absolute (but not relative) weight; of Galileo's principle—which linked the known facts about inclined planes to those of the path assumed by projectiles; and of Watt's principle—which linked the known facts about the compression and expansion of gases (Boyle's law), especially steam, into a meaningful relationship eventuating in the steam engine. Unquestionably the storehouse of facts already catalogued in standard reference sources or available by personal correspondence provides many untapped opportunities for new discoveries and inventions, both of instruments and of ideas. The essential problem at this stage of research, therefore, is twofold: (a) to determine what types of available materials are relevant to the inquiry at hand; and (b) to find such materials.

VI.10 Determining which data are relevant to a particular research problem is an arbitrary decision which each person must make for himself. Avoidance of the extremes seems to be about the only general rule that can be offered. One extreme is that of overloading a study with background material that bears only a superficial relationship to the problem at hand—or, worse yet, of assuming that practically everything ever written on a subject is directly pertinent to the present interest. Much scientistic scholarship exhibits this tendency clearly, just as does much so-called scholarly research (e.g., many Ph.D dissertations). The other extreme is the implication that nothing of any consequence bearing upon the problem at hand has ever been done before. (This is common to pseudo-scientific tracts.) By and large, the question of relevancy can be answered only in terms of the hypothesis and of the total research design. Therefore, each researcher must decide whether certain data are or are not pertinent to his inquiry; but he should be prepared in the event of criticism to be able to defend his choice of materials by demonstrating his knowledge of the existence and implications of those materials not utilized.

B. *Evaluation of Materials*

VI.11 Finding relevant materials is complicated by the fact that relevancy is related to validity. This therefore raises the formidable question of how to assess the validity of published material. A general distinction is often made by scholars between primary (original) and secondary (derived) sources. Actually, however, a primary source will be as biased or as distorted as was the viewpoint of the person who reported it. But it does have an advantage insofar as no further distortion by subsequent reporters can have occurred as could have been the case with secondary sources. Whenever feasible, therefore, primary sources are to be preferred.

VI.12 The validation of published materials, whether primary or secondary, poses some of the most vexing problems in all scientific research. For all practical purposes, published materials may be classified into two groups: those which describe unique phenomena, and those which describe repetitive or reproducible phenomena. The first type are generically classed as histories; the second type have no commonly accepted designation. Since the validity of the second type can be verified by repetition or replication, it is the first type that raises the major methodological problem of validation. In other words, the validity of a reported observation or experiment can easily be checked by reproducing it or by observing it when it is repeated (e.g., as in astronomy, biology, physics, chemistry); but when the reported observation is unique (e.g., an historical event in a culture or in the life of a specific animal or person), then obviously some other type of validation is necessary.

VI.13 To an increasing extent among scientists, validity is being defined in terms of reliability (i.e., consistency or reproducibility); and if an event is unique (and therefore not reproducible), the question obviously arises as to whether it can be validated at all. This, essentially, is the problem the historian faces, whether he is concerned with the history of a person, of an animal, of a plant, or of the world at large. As will be noted later in Chapter X, validation in science is determined by either or both of two criteria: preferably by consistency of reproducibility (i.e., prediction beyond pure chance), or by consensus among authorities. Historical validation, therefore, when it deals with nonreproducible situations, events or processes, can be determined only by the less desirable criterion of consensus.

VI.14 The examination of historical data begins with so-called external or objective criticism. The investigator at this stage is concerned largely with the problem of determining the credibility of the report. In essence, he asks himself whether the report is genuine or spurious, complete or incomplete. To do this he often employs such auxiliary sciences as philology (the study of linguistics), chronology (the study of time sequences), paleography (the study of deciphering ancient writings, their origin, period, etc.), epigraphy (the study of inscriptions, particularly ancient ones), diplomatics (that branch of paleography concerned with ancient official documents), numismatics (the study of coins and medals), lexicography (the study of lexicons or dictionaries), and perhaps even sphragistics (the study of seals). He also may employ physical or chemical tests such as X-ray or infra-red-ray photography, radioactivity, qualitative analysis, etc., to determine the authenticity of a document, painting, artifact, etc. He must, however, be thoroughly informed about the person, event, era, or phenomenon under examination so that he can answer to his own satisfaction the question of the credibility of the reported data.

VI.15 The historian next turns to the so-called internal criticism of his material to round out the picture of relative or absolute authenticity. The analysis at this stage attempts to solve such problems as the actual meaning of a statement—i.e., should it be interpreted literally or figuratively? He wants to know whether specific statements were made honestly—i.e., did the author of the report have any discernible reason for distorting the truth; or was he the victim of personal factors or social forces which could have prevented objective reporting; or perhaps the author reported only some truths or only partial truths. In terms of verifiability, therefore, the historian tries to solve the problem of validity by bringing together all the facts and reasoning ability at his command to establish the veracity of the record being examined. In most cases, the student of historical material is forced to couch his deductions in terms of probability statements of reliability; and these, in most cases, often remain one of the many unknown variables which plague so many studies employing historical methods of scientific inquiry.

VI.16   The final stage in historical analysis is the interpretation to be placed upon all or parts of a report, document, specimen, etc. This final stage is logically independent of the former stages, for at this point the historian moves from description to analysis. As will be noted later, logical analysis, though it rests upon valid description, involves the same deductive processes which are the foundation of scientific method; and accurate description does not in itself guarantee accurate analysis of causal or sequential relationships. Therefore, the cautious student of historical materials always keeps clearly in mind the important distinction between a descriptive report and an historical interpretation. The former is factual, the latter conjectural; and if the function of scientific method is the achievement of reliable knowledge—in other words, to understand in order to predict—then all the rigorous requirements of the analytic method (to be discussed later) must be imposed upon interpreted historical materials before they can be accepted as scientific data.

VI.17   However the questions of relevancy, reliability and validity of source materials are answered, the methodical student will begin the actual implementation of his research design by assembling a bibliography. This compilation will serve two major purposes: (a) to provide a list of sources giving the framework or setting of his problem, and (b) to denote the specific sources which will be utilized directly in the study. It is imperative, therefore, that the bibliography be developed in a systematic manner. Systems vary according to the whims or interests of the researcher, but the essential feature of any functional system is efficiency. This generally requires a cross-indexing system: one set of forms or cards arranged by authors (necessary for assembling the final bibliography alphabetically), the other arranged by essential topics. Thus a given reference may require several cards if several ideas or facts are to be recorded from a single source. Whether the final bibliography should be annotated must also be determined by the researcher. General practice in terms of the kind of study being done usually determines this factor.

C. Instruments and Tools

VI.18   As was suggested at the beginning of this chapter, one of the major reasons for scrutinizing the literature of a given field is that it often suggests which instruments or techniques might be useful in pursuing the problem at hand. The particular choice of appropriate and effective devices for acquiring data may be obvious in the case of some problems, particularly in well established areas of research; but in other cases the choice of particular data-gathering devices often becomes a major problem in itself. In the first place, there may not yet exist adequate instruments or tools appropriate either to the problem at hand or to the kind of data desired. In the second place, the mere fact that certain devices have been employed by previous researchers is not a guarantee that those devices are either appropriate to, or adequate for the data being sought. For both these reasons, the competent scientist needs to have a thorough knowledge of the limits and capabilities of all the significant instruments and tools employed in his area of research. Furthermore, he needs to have the ingenuity to devise adequate instruments or tools appropriate to his methodological problems if the need should arise.

VI.19   A thorough grasp of the literature should assure competence in both the materials and the methodology of the field of interest. The critical researcher will be constantly alert to the seductive pitfall of becoming so intrigued with the paraphernalia of data gathering that he loses sight of his primary objective: viz., acquiring pertinent and verified facts. Particularly in the fields of the social sciences, the accusation that many investigators are overly impressed with, or unduly intrigued by, their instruments seems to have some basis in fact. A critical awareness of the limitations as well as of the merits of the tools and instruments of one's field, therefore, is an important factor in scientific research.

VI.20   In many instances, especially in those fields that are growing rapidly, ingenuity and inventiveness are just as important qualities of research as is encyclopedic knowledge or critical ability. Significant advances in knowledge often are a consequence of the development of new tools or instruments; and the competent researcher constantly strives to (a) utilize the best devices available, (b) modify such devices if necessary to suit his needs, or (c) devise new tools or instruments appropriate to the problem and data at hand. Sometimes ideas for new research devices are suggested by some other field of science, and occasionally modification of existing tools or instruments may help to solve a heretofore baffling problem. Examples of fruitful borrowing and adaptation from one field to another might be illustrated by the case of using radio-active isotopes as tracers (developed by physicists and borrowed by physiologists), of using psycho-physical techniques of measurement (developed by physiologists and borrowed by psychologists), of using psychometric measures (developed by psychologists and borrowed by sociologists), or of using polling techniques (developed by sociologists and borrowed by political scientists). The competent scientist never ceases in his search for more effective tools and instruments relevant to the demands of his problem and to the qualities of his phenomena, bearing always in mind that discovery (of new methods or of new sources of data) is the reward of patient and constant search in often unsuspected fields.

VI.21   Since two essential attributes of scientific method are objectivity and exactness, the growth of modern science is in many ways directly related to the development of objectifying instruments which permit differential discrimination to be expressed in precise and standardized units. The function of instruments, therefore, is to extend the range of the senses, and/or to indicate differences within and among phenomena in terms of standardized quantitative units. Since it is not our purpose here to catalogue all the various instruments used in science, attention will be devoted to the role they play in the area of methodology. At the outset, it should always be borne in mind that instruments never in themselves answer a scientific question; nor, paradoxical as it may seem, do they necessarily insure accurate measurement of real or significant differences. For the significance of a measurement is a function of its role in the hypothesis; and a weak or untestable hypothesis cannot be strengthened by the employment of accurate instruments.

VI.22 Another feature of measuring instruments should be pointed out simply because it is often ignored or perhaps even unrecognized. That is, measurement is in itself one way of defining, describing, or even of knowing phenomena. Distance, for example, cannot be perceived in the abstract—as any space pilot knows only too well, and as any layman can attest who has gazed upon the unmarred surface of the sea—it can only be perceived in terms of measurable units. To say that a certain distance is "far" is to imply that some kind of measurable referent already exists, otherwise the notion of distance cannot be defined or described. "Far," for example, is one thing in reference to a powered missile but a different thing in reference to a snail. In the former case, "far" derives its meaningful definition in terms of thousands of miles per hour as units of measure; in the latter case in terms of inches or feet per hour; but in both cases the concept "far" cannot be conceived outside of some unit of measurement. Furthermore, the contention that something exists though it cannot ever be measured is a semantic paradox; a measure (however crude) is one way of defining or describing the very existence of a phenomenon. The hoary quip of psychologists ("What is intelligence? It is what intelligence tests measure!") contains a basic element of validity. True, there may be other and better definitions of intelligence than those related to the measures obtained on a so-called intelligence test; but it is just as true that *one* definition of intelligence (and perhaps a useful one at that) is exactly what is contended: namely, that which a specific test purportedly measures.

VI.23 This role of measuring instruments is inextricably linked with the problem of definition referred to earlier in Chapter IV. To a large extent the accelerated growth of a given field of science is most often associated with the creation of concepts defined in mensural terms. Conversely, the retardation of some branches of science is most often a consequence of the lack of mensural concepts. To argue, for example, that the "soul," or "will," or "mind," or "values," or "attitudes," or "emotions" exist but cannot ever be measured is to imply either (a) that such concepts are merely verbalisms with no counterpart in objective or empirical reality; or (b) that they have remained vague, amorphous, and possibly even ambiguous simply because no measures have *yet* been devised *and* agreed upon which will define them in terms of objective or empirical referents. Further consideration of this issue would probably interest only a professional logician or philosopher; but for the scientist, however, it remains axiomatic that objective definition and empirical measurement are the interrelated and interdependent foundations of his whole conceptual system of analysis.

VI.24   Observational tools and instruments are too well known to bear extensive listing. The role of magnifiers (microscopes and telescopes), of surveying transits, of optical range finders, of X-rays and fluoroscopes, of spectroscopes, of photometers, rulers, gauges and micrometers is well appreciated by anyone living in a modern culture. Likewise, the role of photographs (both still and moving), often in color and with accompanying sound effects, is being extended constantly to provide both accurate and lasting visual records of both micro and macro phenomena. Without belaboring the point, it would appear obvious that observational instruments and tools have provided one of the greatest aids to the accelerated growth of science.

D. *Special Limitations of Social Science Instruments*

VI.25   But this growth has occurred largely in the physical and biological sciences. In the social sciences observation has been objectified and amplified by some of these same instruments (although not to the same extent), but undoubtedly without comparable results. For in the first place, the social scientist has had particular difficulty in determining *what* to observe as well as *how* to observe it. These deficiencies or limitations inhere in inadequate theory and therefore in relatively infertile hypotheses. In the second place, the relative primitiveness of social science has permitted if not actually encouraged the employment of inaccurate visual tools such as observational schedules. In general use each observational schedule differs from all others. Furthermore, it relies upon the visual acuity and sensitivity of the particular observer. It is not surprising, therefore, that the so-called behavioral scientists (anthropologists, sociologists, psychologists, etc.) have not yet developed a common body of objective, accurate and meaningful data.

VI.26 The lack of a large body of objective, accurate and meaningful data in the behavioral sciences can be illustrated in terms of some specific principles of good schedule construction. These principles serve to synthesize and correlate the various features of clear and functional concepts mentioned earlier (Chapter IV), but also serve to point out some of the common weaknesses of most observational schedules employed for social data. Essentially these principles are: (a) The data to be sought and recorded should be limited to objective and quantifiable phenomena in terms of number, size, age, volume, etc. Data concepts that cannot be objectified or quantified should be avoided in the interests of avoiding ambiguity or inaccuracy—e.g., concepts such as "large," "small," "few," "many," "high," "low," "young," "old," "recent," and others of this vague kind. (b) The data should be limited to types which can be easily and unequivocally observed by any moderately competent observer. Such equivocal concepts as "well adjusted," "poorly adjusted"; "socially acceptable," "antisocial"; "pleasant," "stern"; "intelligent," "stupid"; "cooperative," "uncooperative"; "leader," "follower"; "inner-directed" (introvertive?), "outer-directed" (extrovertive?) —should be scrupulously avoided in the interest of objective precision.

VI.27 Additional principles of this type are: (c) The quantitative units should be limited to manageable sizes so that no guesswork or possible misinterpretation can occur. Particularly confusing are such unit concepts as "average," "rarely," "usually," "often," "most," "few," "never," "always," etc. (d) The data should be limited to phenomena which can be objectively verified through re-observation by other equally competent observers. This would eliminate so-called unique phenomena which cannot be replicated or re-observed by others; or even so-called intrinsic phenomena whose existence is imputed rather than directly observed—e.g., "middle-class environment," "healthy atmosphere," "mass hysteria," "tender loving care," "the spirit of the times."

VI.28   If a code is to be employed to record the observations, (e) a checking system should be used to spot any memory lapses or misinterpretations on the part of the various observers. Schedules specifically (f) should be pretested (1) for reliability, by noting whether the various (or even the same) observers do in fact record the same kinds of facts consistently; and (2) for validity, by noting whether the various observers note and record the actual phenomena designated or implied in the schedule. Though many other essential principles of good schedule construction could be specified, this listing alone should suggest why so many poor studies are done in the various fields of the social sciences.

VI.29   Speech and sound devices, though not as numerous as observational instruments, play the same general role in data gathering as do all other objectifying devices. In the physical sciences the employment of microphones, audiometers and stethoscopes is well recognized, just as is the use of recordings on discs or tapes. In the social sciences, however, apart from the recording of speech or music, the development of objectifying instruments presents innumerable technical problems. Interrogative schedules and interview guides should attempt to standardize both the asking of questions and the recording of verbal responses; but the many uncontrolled variables involved (e.g., timing, voice pitch and inflection, diction and accent, facial expressions, sex, race, age and appearance of the questioner) can significantly influence the quality of the responses. Even under optimal conditions of pretesting such schedules, the language problem (referred to several times previously) often interferes with the collection of an objective, standardized, discretely measurable body of verifiable data.

VI.30 The outstanding interrogative instruments of social scientists are the interview and the questionnaire, and their closely related allies, the rating, ranking and sociometric scales. A rating scale has as its purpose the designation of the relative presence or absence of particular attributes (e.g., intelligence, leadership, prestige, social class), either qualitatively or quantitatively, but usually of a psychological or social order. A ranking scale may be concerned with the same type of attributes or their aggregate form, but its basic purpose is to indicate the relative order (i.e., more or less than) of the persons or phenomena being rated. A sociometric scale has as its purpose the measurement of social distance—i.e., the status and prestige, or relative acceptance or rejection by one's peer groups, or gradations of differences in social behavior. Like the interrogative schedule and the interview guide, questionnaires and scales—even when employed under optimal conditions—contain methodological as well as technical weaknesses not found in physical instruments.

VI.31 To mention just the outstanding problems involved in the construction and use of such devices, the following points should be noted: There are three general classes of items that can be investigated: questions of fact, questions of attitude, and questions of opinion. It is imperative that the investigator know the essential differences between these three classes of phenomena, for each presents particular problems of methodology and interpretation. A response to a question of fact (e.g., "When were you born?" "Where do you live?" "What was your wife's maiden name?") can be checked for validity, provided that some objective and independent source of validation is both existent and available (e.g., birth certificate or other records). An attitude, however, is commonly defined as the mental predisposition to respond to a rather specific situation in a rather consistent manner. (Note already how difficult it might be to define objectively the terms "rather specific" and "rather consistent.") A response to a question of attitude (as defined above), therefore, can be validated only by checking the correspondence between the verbal response and the ensuing related act (e.g., "Who will you vote for in November?"). An opinion, however, is commonly defined as an expression of choice between alternate hypothetical proposals (e.g., "Who do you think will win the election?"). A response to a question of opinion, therefore, cannot be checked for validity in the same sense that can one of fact or of attitude. Since an opinion response is by definition not necessarily related to predictable behavior, there is continued serious question of its usefulness as commonly employed in behavioral research.

VI.32   Ease of administration is likewise a basic problem in the use of interrogative instruments. Some such instruments (e.g., voting polls, consumer-preference scales, traffic checks) are basically simple both in construction and in method of response as well as in administration. Some such devices, however (e.g., opinion, rating, ranking, attitude or "neurotic-inventory" scales), are often quite complex. It is very difficult in many cases, therefore, to feel confident that the results obtained actually are clear and objective indices of the phenomenon being investigated. A fatigue factor also is involved in the employment of interrogative instruments. Unlike a machine, human subjects become increasingly undependable when asked long series of questions. (How long? is a debatable question in itself.) Studies quite clearly indicate that both the reliability and the validity of such devices tend to decrease proportionately to their length.

VI.33   The language or semantic problem also is basic in the employment of all interrogative devices. Ambiguity, stereotyped responses, cliché answers, "loaded" words, etc., plague objective communication. To ask such a question as "What qualities do you desire in a husband?" invariably evokes such stereotyped (and hence undependable) answers as "thoughtfulness," "affection," "intelligence," "maturity," etc. In a well-known nation-wide poll of opinions on class affiliation, over three-quarters of the respondents said that they believed themselves to be members of the "middle" (*vs.* "lower" or "upper") class; but when the item "working class" was added later, this became the most common choice, even for many of those who previously had classified themselves as members of the "upper" class. Granted that people's opinions about such phenomena as desirable spousal qualities or class affiliation may be valuable data—depending upon their use in a particular theoretical framework—it would be most naive to accept such responses as objective facts simply upon their face value.

VI.34 Some interrogative devices depend heavily upon the subject's memory; and memory is a notoriously unreliable source of accurate data. To ask, for example, "Were your relations with your parents very happy, moderately happy, not very happy, etc?" is the height of ambiguity as well as of possible inaccuracy due to unreliable recall. Yet more than one famous study in the behavioral sciences has rested its case upon data no more reliable or valid than these illustrated. As any competent trial lawyer well knows, human memory is notoriously susceptible to suggestion, either by the subject himself due to later experiences, or by other parties purposely intending to sway one's judgment. (This fact has been dramatically illustrated in the case of "brain-washed" prisoners of war.)

VI.35 Another basic problem inherent in interrogative devices is that of possession and knowledge—i.e., does the respondent actually have an opinion on this subject? The previous illustration of preference for certain types of potential spouses is a case in point; but a better example would be the case of so-called occupational-interest scales. Assuming that we now know what personality qualities are necessary for success in a given large number of specific occupations—an assumption, by the way, contrary to fact—it would be highly presumptuous to assume that a subject would be both objective and knowledgeable enough to identify within himself the presence or absence of such specific attributes. (After all, most everyone likes to see himself—and therefore checks such items about himself—as "intelligent," "persevering," "interested in people," "cooperative," etc.) When the question is even more hypothetical than spousal or occupational choice, however (e.g., "When do you think war will break out?"), it is doubtful that any reasonable defense can be made for it on serious scientific grounds.

VI.36 Other essential problems of interrogative devices could also be explicated, but will only be mentioned in passing. One problem is that of scoring: Can the responses be metricized so that they can be treated statistically? (For example: How does one *logically* quantify "very much"?). There is also the problem of weights: are the response units a valid designation of equal-appearing intervals of intensity—i.e., does "very much," for example, have the same potential-response value, or "weight," for all respondents? Finally, there is always the question of whether to employ a "structured" or an "unstructured" device (i.e., one containing limited and predetermined answers *versus* one which allows general and unlimited answers). Both of these two general types have their particular strengths and weaknesses; the former the advantage of simplicity and specificity, the latter (especially in the case of so-called "projective" tests) the advantage of possibly penetrating the subject's resistance or of opening new vistas of analysis.

VI.37 The use of interrogative devices of any kind presents myriad complex problems, many of them still unsolved; and their very complexity requires long and intensive study. Suffice it to say, however, that the basic problems involved in the use of any diagnostic or measuring instrument remain essentially the same: (a) Is the instrument accurate—i.e., does it distinguish clear and meaningful differences? (b) Is it reliable—i.e., does it give consistent results under comparable conditions? (c) Is it valid—i.e., does it measure what it is supposed to measure? (d) Is it practical —i.e., can it be employed under normal conditions of use? (e) Is it pertinent—i.e., are its results relevant to the purposes of the study? (f) Is it functional—i.e., does it give measures that help to answer significant questions? Only when such questions as these can be answered affirmatively can an instrument be said to have methodological value.

E. *Detecting Feeling States*

VI.38 Instruments used to identify or measure sensations are less numerous than are those employed for observation, hearing or language (i.e., for the observation of concrete phenomena or for the designation of verbal responses). The measurement of relatively simple and discrete sensations has, of course, been accomplished for a long time. To mention just a few of the more common instruments, the use of thermometers and of barometers is well known, just as is the use of hydrometers or of hygrometers. The use of the common balance to denote differences in weight is obvious, just as is the use of altimeters to measure elevation, thermocouples to distinguish temperature differences in metals, and the physician's old stand-by, the sphygmobolometer to measure blood pressure. But it is a problem of a different order to combine the various neurological, glandular, respiratory and circulatory reactions of a human in order to ascertain his feeling states or emotions. Of course, such feeling states presumably can be ascertained and measured by verbal methods (as mentioned previously) through interviews, scales and questionnaires; but measurement by nonverbal devices requires another method of approach.

VI.39　The measurement of feeling states or emotions is generally accomplished on the nonverbal level by either of two major methods. The first is by noting the observed behavioral responses of an individual under given conditions, and then—after indicating those responses upon a quantitative descriptive-diagnostic scale—labeling the responses by the name of an emotional condition or reaction. The second method is by noting the internal and external effects of the feelings as measured by a battery of instruments attached to the subject's body. The first method is occasionally employed either in psychiatry or in psychology. Essentially, it consists of observations of predefined actions according to various scales developed by panels of judges. Such actions might be labeled "aggressiveness," "withdrawal," "disoriented," "cooperative," "introverted," "confused," "belligerent," "egocentered," etc., and each action is scored by a system of weights also previously determined. Such scales, however, are not standardized, and are employed to suit the purposes at hand. Since they depend for their validity upon the observer's definition of the behavior being noted, there is no assurance that either the labelled emotional state or its degree of intensity actually are valid indications of the true condition of the subject. For this reason, such devices are employed only in such special situations where standardized verbal scales (either of the interview type or of the paper-and-pencil type) are inapplicable. Such scales, incidentally, do not have comparable validity to nonverbal "intelligence" tests (i.e., those requiring the performance of specific manual tasks)—poor as the latter are. It seems highly possible, however, that ingenuity by psychologists will eventually develop far better emotional scales of the nonverbal type than any now existing.

VI.40   The second method of measuring emotions employs a complex instrument—the pneumopsychopolygraph, or polygraph for short (popularly called a lie detector, which it is not)—which records on paper tape such reactions as breathing rate, carbon dioxide content of the breath, body temperature, pulse rate, acidity of the skin, changes in the electrical potential of the brain (like an electroencephalograph), heart contractions (by an electrocardiograph), etc., in relation to the questions being asked of the subject. To the extent that an emotion can be defined as a change in feeling states, so then can such an instrument serve many useful purposes. But if an emotion is defined as the combination of an attitude and its accompanying feeling states, then such an instrument indicates at best only one half of the total reactional complex. It is for this reason that opposite emotions give comparable results on such instruments, for only the presence and the intensity of the internal responses are measured —not the stimuli (i.e., the attitudes) provoking those responses. Thus, for example, asking the questions "Do you like to eat snakes?" and "Do you like to eat roast beef?" might produce responses of comparable intensity; although in the first case the contributory stimulus would be a negative attitude (i.e., fear or revulsion), while in the latter case the stimulus would be a positive attitude (i.e., pleasurable anticipation).

VI.41   It should be obvious, therefore, why so-called lie detectors have limited applications in the measurement of emotions. They are generally reliable in distinguishing the presence of a response-set in most subjects—but not in all. For example: If an interrogator should ask a suspect: "Did you rob the store?" the respondent obviously would react strongly even if he were innocent, for he would know that the question is significantly related to his future welfare. A "strong" response in such a case would be uninterpretable, except to indicate in quantitative terms the degree and duration of the subject's excitation. If the interrogator should ask, however, "Where is your pocket knife?" then the measured response could be meaningful; for, to pursue the example, it may have been established that a pocket knife related to the criminal act had been found at the scene of the crime, and only the police and the thief could possibly know this fact. Yet some persons apparently are able either to control or to randomize their emotional responses to such an extent that the readings of such instruments are meaningless to the interrogator in terms of the questions provoking the responses. It also often happens that a suspect (not necessarily guilty of anything) is so upset by the ominous-appearing machine and the legal "climate" within which it is employed that his recorded patterns of responses are completely erratic—much more so than they should be for a normal innocent person—due to his apprehensiveness alone. It is for these reasons that the results of such instruments are not generally permitted as evidence in court trials.

VI.42 A word should be said at this point about the use of photography in the measurement of feeling states. By and large, this method is highly unreliable. Photographs of subjects under emotional stress permit a wide variety of interpretations. Observers are notorious for misinterpreting facial or bodily expressions, just as they are notoriously unreliable when asked to judge a person's "character," occupation, age, interests or temperament from his photograph. Popular fiction to the contrary, neither "character" nor emotion is more than coincidentally related to physiognomy, body types, race, sex, nationality, etc., and the use of the camera generally tends to suggest only stereotyped responses. Since the visible features of a person's responses bear no necessary relation to his feeling states (as any good actor knows), objectively recording those features has a very limited value for the scientific psychologist.

VI.43 The last general group of sensations, tasting and smelling, have hardly been studied at all in comparison to the other senses. By and large, both the presence and the intensity of oral-olfactory sensations have been designated only in very general terms. Even the single instrument sometimes employed to denote olfactory "thresholds" (i.e., least perceptible differences), the olfactometer, is extremely limited both in range and sensitivity. At the present time, the presence and intensity of taste or smell stimuli-effects remains a crude and highly subjective field of scientific designation.

VI.44 A miscellaneous group of instruments are utilized by scientists to extend the various senses, but have not been mentioned previously due to either their obvious or highly specialized character. In this group may be mentioned the clock and the chronometer to measure time, the anemometer or aerovane to measure wind speed and direction, sonar and radar to measure distance and direction, the manometer or ergometer to measure muscular effort involved in work, etc. All these exact instruments are employed in various phases of scientific research; and both the rationale underlying their use and the utilization of the data they measure are obvious to the specialists who employ them. It is important to bear in mind, however, that even exact instruments are not foolproof. The cautious scientist checks his instruments before they are put into use, and then checks them periodically against possible errors requiring recalibration. Furthermore, he standardizes his interpretations of the readings in order to minimize inconsistency. Finally, he seeks constantly to improvise or invent even better tools and instruments in order to increase the exactness and sensitivity of his measurements; for he soon learns from experience that to a certain extent he is bound by their limitations just as much as he is bound by his hypotheses or theories.

VI.45 A final word related to measuring instruments should be added at this point about the role of automatic data-processing machines. Unlike ordinary adding machines or calculators, electronic computers add a new dimension to both the theoretical and mechanical aspects of scientific development. Superficially they permit the more rapid processing of complex data; but much more importantly, they permit the employment of much more elaborate calculations than were previously possible. In a sense, their role may be likened to the invention of the calculus; which, though it did not add any new data to science, nevertheless opened an entirely new dimension to human manipulation of ideas. Electronic computers are calculators, of course, but of such previously unachievable efficiency that they now permit within a few minutes the processing of data so complex that a comparable task formerly would have required the combined efforts of scores of mathematicians working for years. In this sense of efficiency, then, it may be said that such machines now permit the solution of problems heretofore insoluble. Work in factor analysis, for example, now possible with such machines, has changed much theoretical thinking drastically. The recent spectacular achievements in physics and mechanics attests to the vast intellectual "breakthrough" which such machines have made possible. Their potentially significant employment in the social sciences, however, awaits the development of more rigorous (i.e., exact and quantifiable) concepts and formulations.

VI.46   In summary, then, a thorough analysis of the literature and a thorough knowledge of the various tools and instruments of a given field can be seen to be critical factors which often spell the differences between the success or failure of a given research project. Examples are numerous of research novices who were ready to rush into print with startling discoveries only to find that the apparently brilliant ideas were already quite familiar to anyone acquainted with the literature. Examples are also numerous of research novices who marvelled at their own ingenuity at devising or exploiting a particular tool or instrument only to be embarrassingly criticized later for a naivete that could have been avoided by proper preparation through a study of the available literature. Good science rests upon sound scholarship, and growth in science cannot proceed ahead of the tools and instruments available. In the chapter to follow the importance of these two phases of scientific effort will be clearly illustrated in a most significant stage of all science: viz., gathering meaningful and verified data.

CHAPTER VII

Data Collection: Problems of Methodology

A. DETERMINATION OF TECHNIQUES

    1. *Which techniques to employ*
    2. *Understanding their essential features*
    3. *Quantity not related to validity*
    4. *An example of a "scatter-gun" approach*
    5. *An eclectic approach*
    6. *The two basic functions of all methods*

B. OBSERVATION

    7. *Facts* versus *their meanings*
    8. *The role of the senses*
    9. *The process of data gathering*
    10. *Observation*
    11. *Basic questions of observation*
    12. *Developing observational awareness*
    13. *The role of insight*
    14. *Developing objectivity*
    15. *Problems of doubtful judgment*
    16. *Objectifying observations*
    17. *Developing visual clarity*
    18. *Types of objectifying aids*
    19. *Reliability and validity*
    20. *Four types of observation*
    21. *Participant observation*
    22. *Essential features of good observation*
    23. *Observational pertinency*

C. HISTORICAL DATA

D. MEASUREMENT

E. RECORDING AND PRESENTATION

A. *Determination of Techniques*

VII.1 A major decision in any research design is the determination of the techniques to be employed for gathering data. In practice this decision may have been made before the bibliographical phase of the study had actually been initiated; and the two stages—library research and original research—might be carried on simultaneously. In some cases the characteristics of the problem rather clearly indicate the technique to be employed for gathering original data: e.g., the experiment in chemistry, excavation in archeology, photographs in astronomy, or the case history in psychiatry. In other cases, however, the researcher will need to decide which technique or combination of techniques seems to offer the most fruitful possibilities of answering the questions posed by the hypothesis.

VII.2 It is imperative, therefore, that the competent scientist understand clearly the essential features of various research techniques in terms of their particular strengths or weaknesses. No single technique is adequate for all problems; nor is any particular technique inherently more "scientific" (i.e., accurate) than any other. But neither is the employment of all possible techniques necessarily superior to the employment of any single technique in order to promote reliability or validity.

VII.3 The last point deserves elaboration, for there is a too-common tendency in some fields (particularly the behavioral sciences) to equate volume and variety with validity. Or, put another way, there is a common tendency to imply that the massing of data derived from different techniques is more impressive —i.e., more convincing because presumably more reliable—than would be the case from data derived from a single method alone. (In spite of this writer's position that the scientific approach can most accurately be described as *a* method, he temporarily bows to common usage; and hereinafter will employ the two terms "method(s)" and "technique(s)" interchangeably.) This implication is essentially a "card stacking" device of persuasion if, as is often the case, there is no logical or methodological reason to presume that sheer quantity or diversity necessarily promotes validity.

VII.4 A common example of this "scatter-gun" approach would be the explanation of an individual's anti-social behavior. In the attempt to "explain" such behavior, a researcher may offer (a) a medical history, (b) a social history, (c) an educational history, (d) intelligence-test scores, (e) opinions from truant officers, teachers, parents, neighbors, friends, etc., (f) attitude-test results, (g) psychiatric-interview opinions, etc. Now, it may be that such behavior can be understood and hence explained only in terms of multiple factors of causation; and that each of such factors can only be determined by the employment of specific and essentially different devices. But it is also possible that such an approach simply beclouds the determination of causative factors by diffusing to the point of indeterminability any possibility of "pinning down" the essential factors which are actually responsible for such behavior.

VII.5 This problem of selection (of single versus multiple techniques) should not be confused with the frame-of-reference issue to be discussed later (see Chapter IX, D). If a researcher's frame of reference dictates an "eclectic" approach, this implies that he believes that different portions of different theories are equally valid. And if he is convinced that, according to his hypothesis, causation is due to multiple rather than to single causes, then obviously he should be searching in different areas for those various causes. But the search for multiple and therefore interdependent causes does not necessarily demand the employment of essentially different techniques. As in the example above, the determination of which technique(s) is logically related to the inferences of the hypothesis is a question not solved simply by employing any and all techniques available. The problem of pertinency (i.e., logical relevance) remains a separate issue.

VII.6 Any specific method may be more appropriate, more efficient, or more convincing than others in a given situation, depending upon the type of problem posed, the class of phenomenon to be investigated, or the time, money, manpower or facilities available. All methods, however, serve either or both of two basic functions: namely, (a) to answer the descriptive question "What is the fact?" or (b) to answer the analytic question "What is the relation between?" Whether the answer to either question is to be sought in qualitative or in quantitative terms, selection of the most fruitful techniques and the rationale for their choice are factors which the investigator should understand clearly. Otherwise he can never be sure why he gets the results he does, what the results actually "prove" (if anything), or how he might improve his efforts in the future. When employing particular techniques within an analytic design, however, specific problems of inference present themselves which are not normally found in purely descriptive studies. The following discussion, therefore, will be directed toward an examination of the basic structural features inherent in analytic designs.

B. Observation

VII.7 A basic postulate of scientific method is that all data are derived from sensory impressions. This does not deny, however, that impressions may be purely mental, for one way of defining thought is as the mental manipulation of sensory impressions. The data of science, therefore, from this perspective, are mental impressions of sensory experiences—i.e., ideas derived from seeing, hearing, smelling, tasting, touching, etc. Whether the impressions or sensations are directly or indirectly derived does not alter the essential fact that man reacts to the mental manipulations of sensory impressions, and these reactions, or ideas, are the foundation of knowledge. Facts do not "speak" for themselves; only the "meaning" of those facts makes an impression on the mind's awareness.

VII.8 Perhaps it should be pointed out that speaking in terms of the five senses is simply a convenient and popular way of denoting man's perceptive abilities. Actually, there are a number of well recognized senses, not just five. The kinesthetic sense, for example, orients us in space and enables us to control the movements of our bodies. Other senses give us impressions from our inner organs, yielding feelings of hunger, thirst, nausea, pain, etc. Furthermore, most senses are not simple entities; rather, they usually comprise several related sense faculties. Touch, for example, involves several sense-discriminatory abilities, including texture, pressure, pain, temperature, etc; while both hearing and seeing differentiate the quality as well as the quantity of impressions. A common method of classifying the senses may help to suggest the actual complexity of the total sensory mechanism of humans by indicating the three general groups involved: (a) proprioceptive senses, which gives us knowledge of the body itself (through muscles, tendons, joints, etc.), (b) interoceptive senses, which give us knowledge of conditions inside the body (hunger, stomach pains, bowel and bladder pressure, etc.), and (c) exteroceptive senses, which give us knowledge of outside stimuli (texture, temperature, humidity, air pressure, etc.).

VII.9 Data gathering in science, therefore, is essentially a process of (a) receiving stimuli from the phenomena being studied; (b) manipulating the impressions of those stimuli mentally in order to interpret them (i.e., arrange them into meaningful relationships); (c) combining those impressions with other previous impressions and their interpretations (i.e., with memory); and (d) deducing therefrom a concluding interpretation of the phenomena. Two elements are involved in this process: the sensory organs (or extensions of them such as instruments), and the mind. Which sense organs will be employed obviously depends upon the properties of the phenomena being studied; and the mental manipulations will be related to the knowledge, mental ability, awareness, interests and attitudes of the person involved. In the majority of cases, only three of the five well-known senses (seeing, hearing and touching) and speaking are employed in the data-gathering process; and the majority of methods of science employ only two of the five methods of communicating those impressions: viz., observation and interrogation.

VII.10 The basic method of data gathering in science is observation. Whether the scientist looks at a lump of coal, at the stars, at an animal, at a plant, or at other human beings—and whether he looks directly or through a visual accessory such as a telescope or a microscope—observation is by far the most commonly-employed method of ascertaining what *is*. For this reason, the competent scientist trains himself to observe as accurately as possible, first by developing a sensitivity to things that are pertinent to his interests, second by developing as unbiased an attitude as possible toward his phenomena, and third by employing various kinds of visual aids which assist him to clarify what he sees.

VII.11   Before the trained observer can proceed, however, several basic questions relevant to any scientific observation must be answered: (a) What phenomena are to be viewed? Which behaviors are to be selected from the total mass of possible phenomena? Which seem to be the meaningful facts to be looked for in the welter of often confusing data? In short, which observations are pertinent? The answer to these questions can be found only in the hypothesis. (b) Under what conditions should the observations be made? At what times? In what areas? How is the observed situation to be portrayed or structured (i.e., arranged)? Again only the hypothesis can provide a logically meaningful framework for delineating the appropriate observation. (c) Is there any evidence (and if so, of what kind) that the observed phenomena are composed of functional units or processes which are demonstrably *related* to the hypothesis? (d) Can the observation be quantified or metricized in standardized units according to a valid and reliable scale of measurement? (e) Can the conditions of the observation be standardized and stabilized to permit verification? Only when all these questions can be answered positively can one feel that his observations are truly scientific in the complete sense of the term.

VII.12   The initial stage in the development of observational accuracy is the employment of sensitivity, or awareness. This quality is partly a consequence of experience and partly a consequence of insight. Experience, of course, implies directed and not random observation. The experienced observer is one who learns to look for particular things to the exclusion of other, irrelevant things, and at things in a particular way according to

his purposes. In science, observational experience suggests looking at certain qualities or quantities as they appear to be, not as
they are supposed to be or as one would wish them to be—in
short, objectively.

VII.13 Things are viewed by a competent scientist, first, according to their empirical properties, and second according to a
reasoned interpretation of what they should look like as interpreted by other experienced observers. Experience alone, however, is not enough to insure accuracy, for even the most experienced observers may occasionally be proved wrong. Insight, or
the ability to "see through" and beyond the obvious attributes
of a phenomenon connotes a particular type of observational
awareness. Essentially, *insight* refers to the ability to see qualities
or relationships not evident to most observers; and deriving an
insight is largely a matter of intellectual ability—particularly the
kind of ability defined by the Gestalt school of psychology as
intelligence or, more specifically, "structural thinking." Observational sensitivity, then, is a combination of trained experience
and insight.

VII.14 The second stage in the development of observational
accuracy, objectivity, or lack of bias, is relatively easier to achieve
in the physical than in the social sciences, simply because the
observer cannot identify himself with his phenomena. Yet the
history of physical science is replete with instances of inaccurate
observations even in the case of such impersonal phenomena as
the sun, the earth, fire, plants, gravity, force, energy, etc. The
reason for such distorted observations is obvious: although such
phenomena are impersonal, man nevertheless is vitally related
to them by their effects upon him; and therefore often finds it
difficult to view them with an unbiased attitude.

VII.15 A basic and significant datum of experimental psychology is that one sees what he wishes to see or has been led to
expect; conversely, one does not see what he wishes not to see or
has been led to deny. The same fact applies to other sense qualities: hearing, feeling, tasting, smelling. Aware of the implications
of this fact, the scientist trains himself assiduously to question
his observations in order to counteract the possible distortion

due to bias or "wishful thinking." When he does not feel reasonably confident that his attitude toward his phenomena is unbiased, he checks his observations with others who are assumedly unbiased or who at least possess different biases. In cases of doubt, the pooled judgment of various observers is assumed to be the least biased observation possible under the circumstances.

VII.16 The use of assumedly unbiased or of differing-bias observers as a check upon one's observations does not necessarily insure validity, however, for the history of science demonstrates quite clearly that all the assumedly competent observers at a given time were later proved to be wrong. But in the main, this problem of biased observation today occurs far more often in the social than in the physical sciences. This is not, as is often erroneously assumed, the same problem as objectification of observation. An observation may be objectified by recording it with an impersonal instrument such as a camera. But the essential problem is to see things as they actually are; and no camera can objectify subjective phenomena such as hate, love, intelligence or sin; it can only record what the photographer places in front of its lens as an example of the thing that *he* "sees." The validity of an observation, therefore—as contrasted to its objectification—in the last analysis becomes part of the larger problem of verification, a problem to be discussed in detail in a later chapter.

VII.17 The third stage in the development of observational accuracy is the employment of visual aids in order to clarify (if not also to objectify) what one sees. In common experience eyeglasses, telescopes or microscopes are obvious clarifying aids to vision; and where the phenomena to be observed are composed simply of discrete, physical properties, the standard aids usually insure observational accuracy. But the clarification of other types of observation is not necessarily so simple as the above remarks might suggest. Microscopes or cameras depict what is placed in their field of view, and convey an image lighted in a particular way. The essential quality to be observed may not be observable because (to pursue the analogy) the microscope or camera viewed the wrong scene, or at the wrong angle, or in the wrong light, or at the wrong time. In summary, then, it is important to bear in mind that visual aids assist only in solving the physical problem of seeing the phenomenon; they cannot solve the problem of seeing the "right" phenomenon or in the "right" way. (What constitutes "right," of course, is determined by the theoretical structure of the problem.)

VII.18 Objectifying aids are of three general types: (a) those which project the senses or increase their power (telescopes, microscopes, sound amplifiers, etc.); (b) those which indicate more discrete units of differences than can be felt, heard, smelled or observed (calipers, graduates, gauges, scales, etc.); and (c) those which combine these two functions of projecting and increasing the senses and indicating discrete measurable differences objectively (thermometers, barometers, anemometers, psychogalvanometers etc.). The accuracy of any of such instruments depends upon two independent qualities: (1) their reliability, or ability to give the same measure under repeated applications (assuming the conditions have not changed in the interim); and (2) their validity, or ability to indicate the actual properties they are presumed to measure. Any scale or instrument is reliable, therefore, if it is consistent under comparable conditions; but it is valid only when it gives the same results as does another already validated instrument of the same kind when measuring the same phenomenon.

VII.19 Achieving reliability or consistency of designation is generally a relatively simple problem capable of solution by relatively simple means (whether of physical or verbal scales). Ascertaining validity, on the other hand, at times presents almost insurmountable problems. In the physical sciences, "master" scales (e.g., the international meter bar in Paris) have been established by which the validity of any standard measuring instrument can be checked. But in the social sciences such "masters" do not exist. The problem of validating an "attitude" scale, a "temperament" test, or an "intelligence" test, etc., is practically insurmountable if one should insist upon a master referent of validity. Increasingly, therefore, the notion of validity in the use of social science instruments and scales is being replaced by the more defensible concept of reliability. This question is of such paramount importance in all branches of science that it will be dealt with more thoroughly in Chapter X.

VII.20   Observation may be any one or a combination of four kinds: uncontrolled-nonparticipating, uncontrolled-participating, controlled-nonparticipating, or controlled-participating. Each kind has its particular advantages and disadvantages, depending upon the significant qualities of the situation and the purposes of the observer. Uncontrolled observation is particularly appropriate for phenomena which cannot presently be controlled (e.g., stars, tides, death, aging), or should not be controlled because, in the demands of the hypothesis, to do so would be to alter their essential attributes (e.g., erosion patterns, animal nesting habits, leadership responses in group relations, consumer habits). On the other hand, controlled observation may be either necessary or desirable, depending upon the demands of the hypothesis, in order to facilitate an experimental design; provided, of course, that controlling the phenomena does not alter them significantly for the purposes inherent in the research design.

VII.21   Participation in the field being studied likewise is impossible in some cases (e.g., astronomy, ornithology, archeology, insanity), or undesirable in others because involvement in the phenomena (e.g., effects of disease, widowhood, prejudice, divorce) could lead to a distortion or to a limitation of one's perceptive abilities. On the other hand, participation may be either necessary or highly desirable in those cases where only direct experience would permit the kind of observation necessary to develop insight into the essential qualities of the phenomena being studied (e.g., effects of altitude upon pilot fatigue, of prolonged fasting upon reasoning ability, of minority group membership upon voting attitudes, or of parenthood upon child-rearing attitudes).

VII.22   In summary, it should be borne in mind that observation is the basic and most commonly-employed technique of scientific study. The essential features of good observation are accuracy, pertinency, insightfulness, reliability and validity. Training in observation is a *sine qua non* of any scientist; and learning to use instruments of measurement (and whenever possible to develop better ones), is fundamental to the gathering of accurate data. It should never be forgotten, however, that attaining accuracy, reliability and validity is only one-half of the observational problem, the other half is the pertinency of the observation. It is often in this respect of looking for the pertinent qualities that the genius is distinguished from the mediocre scientist; for the difference between these two is the difference of insightfulness—a quality which no accurate, reliable or even valid instrument can bestow.

VII.23   Pertinency of observation often plays a spectacular role in science by inadvertency or chance. Many of the famous discoveries basic to modern science have resulted from accidents or errors in design or in method. Yet the point that is generally overlooked in this respect is the fact that the accidental discovery was made by a scientifically oriented person who was (a) alert enough to, and (b) informed enough of, the potentiality of the discovery that he was able first to see it, and secondly to realize its significance when he saw it. (Among the more spectacular and highly significant discoveries of this sort might be mentioned the principle of buoyancy by Archimedes, the galvanic reflex by Galvani, the cause of childbed fever by Semmelweis, the discovery of penicillin by Fleming, of immunology by Pasteur, and of antisepsis by Lister.) No scientist has ever made a significant discovery simply by looking about randomly or at meaningless facts. Rather, such discoveries were made by trained observers looking for specific relationships inherent in a mass of pertinent facts ready to be fitted into the context of a meaningful hypothesis.

C. Historical Data

VII.24   Histories, like interviews, polls or questionnaires, need to be viewed for what they are: interrogative methods. Whether the history is of a single person (like the case history of the psychiatrist) or of a whole nation, and whether it is derived orally in a face-to-face situation or from modern or ancient written records, its scientific function is the same: to answer specific questions of fact posed by the hypothesis. A so-called general history is of no more value to a scientist than is any other random conglomeration of facts. To be useful for scientific purposes, a history should contain accessible facts that are pertinent to a meaningful hypothesis; and of such types that they can be checked for reliability and validity. As was noted in the previous chapter, few histories meet all these qualifications.

VII.25   The case history seemingly presents a paradox to the investigator of medical, psychological or social relationships. On the one hand, it is patently true that the present is a consequence of the past; therefore to understand the present one must ascertain the particular antecedent or past phenomena which resulted in the present. A stomach ache, for example, is usually the consequence of what one has eaten in the immediate past, or perhaps a consequence of organic disorders developing for some time. Likewise, a phobia is a consequence of past fear experiences; and race prejudice is a consequence of particular sociopsychological events which transpired in the history of a particular culture. In short, whenever a phenomenon is a consequence of a particular series of events, its explanation can be found only in its particular antecedents, i.e., in its history.

VII.26   On the other hand, the wealth of detail usually available in a history encourages a biased investigator to find practically any answer which his predilections convince him he should, and therefore will find. In the case of investigators limited in their analytical thinking by "single-factor" theories—which explain all consequences as the result of single causes—the case history is bound to provide the "right" answers even in the face of what, according to another theory, would be contrary evidence. Psychoanalytic case histories demonstrate this predilection clearly, as do histories which "prove" the racial basis of intelligence, the climatic basis of morality, the hormonal basis of temperament, the genetic basis of criminality, etc. The widespread popularity of the case history, therefore, as a combined descriptive and analytic method of data gathering lies in both its virtue and its vice: the virtue of providing data not ascertainable otherwise, and the vice of encouraging verification of biased perception.

VII.27 A scientifically useful history, therefore, must first of all contain facts that are accessible to the investigator, and many histories are deficient simply in this respect of incompleteness. Accessibility, however, is directly related to the second factor of a useful history, namely, pertinency; and the pertinency of the facts found in a history can be determined only by the requirements of the hypothesis. In this sense the hypothesis serves as the screen which determines which accessible facts are relevant and which are irrelevant (however interesting they may be in themselves). Thirdly, a useful history should be composed of facts which are reliable; that is, which are consistent within themselves and with other interrelated facts of demonstrated reliability. This problem is basically a logical one—i.e., it involves both the semantics and logic of internal consistency— unless the facts are treated statistically (which they rarely are in histories). Finally, a useful history should employ only facts which are verifiable. For all practical purposes the verification of a historical fact is accomplished by checking it against another related and already verified fact. Thus, for example, if a patient should say that he ate green apples yesterday, that historical fact could be verified by recourse to some objective evidence existing independently of the patient's own memory. Obviously in many instances objective verification of a historical fact may prove to be well-nigh impossible, either because there is no other available source of verification or because all the other available sources are inconsistent with each other.

VII.28 Historical validation, however, as contrasted with the verification of a specific event in a particular history of a person or a group, is quite another scientific problem. The former (the validation of a history), means in effect to prove in terms of probability prediction that a given historical interpretation is correct. In this sense a historical interpretation is subject to the same rules of logical inference that are employed in all analytic designs (to be discussed in the next chapter). When a historian feels that he has obtained a meaningful history, he treats the data in essentially the same manner as does the experimenter: by viewing some facts as constants, others as variables, and a third group as dynamic or causative agents. If his hypothesis is to be confirmed, he should be able to make a prediction of consequences expressed in a probability statement—as, for example, when the physician says: "If you continue to eat green apples I predict that you will get a stomach ache within the next three days." Social histories, in contrast to personal medical histories, do not yet permit such a high degree of predictive accuracy; while in the hands of a skillful psychotherapist, psychological histories lie somewhere between these two extremes.

VII.29   Since histories of social events rarely if ever follow the classical "experimental" design, the question obviously arises as to their scientific value. Since the subject of causation or explanation is treated at length in the next chapter, only a few remarks about the development of a "scientific history" or "science of history" need be made at this point. History as an intellectual discipline can be justified on its own grounds, just as can any other intellectual endeavor, without any inherent relation to scientific objectives. In short, persons may study history for any number of obviously legitimate purposes (e.g., for sheer interest alone) without feeling any need to answer, for the querying scientist, the question "So what?"

VII.30   But when history attempts to function as a true social science, then it is obligated to explain events in terms of their demonstrably necessary antecedents or consequents. In this endeavor it is subject to the same logical demands as is any other science concerned with ordinary time-space relationships. To the extent that its explanations are largely "genetic"—i.e., derived in terms of necessary or sufficient prior conditions—then to that extent is it expected to follow the analytic design to be explained in the following chapters.

D. Measurement

VII.31   A fundamental decision in any scientific effort involves the problem of measurement. This problem is twofold: what to measure, and how to measure it. The decision made in answering this dual problem most often spells the difference between mediocre and first-rate research. The following section attempts to unify several strands of the total scientific effort which have been dealt with in previous sections of this book.

VII.32 The first phase of the problem, what to measure, entails questions of conceptualization, classification (taxonomy), validity and pertinency. Specifically, the problem of what to measure demands that some basic decisions be made regarding the specification of units to be employed in the research design. Now, the purpose of an analytic research design (to be examined in the following chapter) is to test hypotheses which have been developed at random or deduced from particular theoretical notions ("laws," "principles," "theories," etc.). However, the procedure involved in moving from theoretical (and therefore abstract) notions to empirically testable propositions (i.e., hypotheses) derived from such notions is not a direct one. Herein lies a major hazard in much ill-conceived research.

VII.33 This hazard is inherent in the fact that theoretical concepts can only be logically defined in terms of certain so-called "primitive" terms—i.e., terms which are assumed to be understood because they cannot be more specifically defined. This would be the case with the term "line" or "point" in geometry. Agreeing that one understands what these terms mean, it now becomes possible to define a "circle," a "triangle" or an "octagon" by reference to "lines" or "points." Theoretical concepts, therefore, can be logically defined only in terms of other theoretical (i.e., abstract) concepts.

VII.34 Measurement, however, demands an operation, and hence can be employed only with operational definitions. The initial problem, therefore, requires the translation of theoretical concepts into operational equivalents. Therein lies the major hazard referred to earlier, for there is *no* purely *logical* way of doing this. Achieving a satisfactory operational definition of a theoretical concept is a matter of *consensus,* not of demonstrable logical equivalence. And achieving consenses can lead to interminable and often indeterminable debate among perfectly sincere, objective and equally competent scientists.

VII.35 An illustration of this fundamental problem can be drawn from common experience. Suppose, for example, that a theory contends that political conservatism is positively related to income. A test of this theory would attempt to demonstrate that persons of high income are politically more conservative than are those of low income. (The problem and the statement of the hypothesis are herein being over-simplified for purposes of illustration). The difficulty, now, is simply and exactly that of achieving agreement on an operational equivalent of political conservatism. Obviously, there may be several equally useful and cogent versions of such operationally-equivalent definitions; but there is no *logical* way of determining which particular definition is more *valid* than another.

VII.36   Ideally, the various definitions contended to be the operational equivalents of a theoretical notion should all give the same results when employed in measuring the notion referred to. In practice, however, this rarely happens; and therefore it is commonly argued that different operational definitions (of presumably the same thing, of course) actually measure different phenomena. This would imply, in effect, that where more than one operational definition for a theoretical concept is employed, then more than one hypothesis is actually being tested. And if more than one hypothesis is being tested, it is certainly highly questionable that the concept used as the major referent (e.g., political conservatism) is being tested or demonstrated at all. It may be, of course, that a given concept has several facets or dimensions, not just one; and that therefore the different hypotheses expressed through the different operational definitions are simply indicative of the sub-parts of the whole. This problem, however, can only be referred to the original discussion of conceptualization in Chapter IV.

VII.37   The second problem, how to measure, is correlated to the first: what to measure. Recalling earlier discussions of conceptualization (especially in Chapter IV), the problem now is how to *order* the presumably *already classified* data. Classification itself, however, is a form of measurement. Specifically, classification involves the ordering or arranging of phenomena into a *nominal scale*—i.e., a scale which separates units in terms of certain discreet and specified attributes. Such attributes can be any quality deemed to be significant, but they must be clearly distinguishable and separable. In our test of political conservatism and its relation to income, for example, we might distinguish various kinds of political conservatists on a nominal scale according to such attributes as race (ignoring the problem of classifying "half breeds"), religion, sex, parenthood, etc. It is vital in any nominal scale that the classes designated do not permit a subject to be placed in more than one category.

VII.38 A nominal scale carries with it *no* implications of weight (i.e., relative amount of a quality—e.g., young *versus* old) or relationship (e.g., better or worse than); but the categories must be exhaustive (i.e., include all cases of a given kind), and yet exclusive (i.e, permit no overlapping). Such a scale, in spite of its limitations, nevertheless serves several useful purposes. In the first place, it permits one to make inferences of "symmetry." This means, for example, that a relation which holds between X and Y also holds between Y and X. In the second place, it permits an inference of "transitivity." This means, for example, that if X is equal to Y, and Y is equal to Z, then X is equal to Z. Combined, these two types of inference permit one to make cross inferences about relations among the various classes of phenomena by employing them interchangeably. It should be clearly noted, however, that since a nominal scale simply classifies but does not infer degree or quantity, the various classes cannot be manipulated mathematically (e.g., by adding or subtracting numerical equivalents of those classes).

VII.39 In order to designate the *relative degree* (but *not* the *absolute amount*) of an entity, an "ordinal" scale is required. An *ordinal scale* designates the relative position of classes in relation to each other. Thus an ordinal scale of political conservatism, for example, might designate persons in terms of such relative degrees as "very conservative," "conservative," "unconservative," "very unconservative," without specifying how much conservatism is indicated by each class. In short, it designates the rank order of sub-classes of a given phenomenon.

VII.40 The superiority of an ordinal over a nominal scale inheres in the fact that an ordinal scale has not only the property of symmetry but also of asymmetry. This means that classes may be designated not only as equivalents to other classes, but also as nonequivalents. Thus, for example, a given ordinal scale may designate that class X is greater than class Y, and therefore that class Y is smaller or less than class X. Transitivity, of course, still operates in an ordinal scale; for if class X is greater or higher than class Z, then any specific X is greater or higher than any specific Z.

VII.41 If it is desired to indicate *how much* one class of a scale differs from another (e.g., how much greater or higher X is than Y), then an *interval scale* is required. Measurement, in the common sense of the term, normally begins at this level of designation. The new element added to this stage is the *unit* of measurement (feet, years, degrees Fahrenheit, etc.). The units so employed, however, should be *constant* and *replicable*—i.e., they should give consistent readings under equivalent measurements.

VII.42 In order to permit such equivalent designations, such units must be located with reference to an absolute or arbitrary zero point. When so located, the scale is now referred to as a "ratio" scale; and has become the most useful type of all measurements. An illustration of both the interval and the ratio types will perhaps help to point out their relative merits. Suppose, for example, that in our test of political conservatism, one person is designated as "ten degrees" conservative, and another as "twenty degrees" conservative. In an interval scale, there is no implication that the first person is in behavioral (i.e., operational) fact only half as conservative as is the second, nor that the second is twice as conservative as is the first; but only that the first is ten degrees—whatever that may mean!—less conservative than the second, and vice versa. Manipulating units in an interval scale by normal mathematical procedures is a wholly indefensible operation common to naive methodologists.

VII.43 If, however (to pursue our illustration), we were able to establish an absolute or nonarbitrary zero point of political conservatism—equivalent in this case to, let us say, a zero amount of money in the bank—then such a scale would move from an interval to a ratio type; and would permit all the usual mathematical and statistical calculations necessary to achieve exacting specification and prediction. It would be possible, for example, to indicate that a person who scores four hundred on a ratio scale of political conservatism is (i.e., predictably, behaviorally speaking) eight times as conservative as one who scores only fifty. Such designations—it should be pointed out, in spite of our illustration—are not possible at this stage of development in the behavioral sciences, though of course they are quite basic and common to the physical sciences.

VII.44   It should be obvious, then, why measurement is so critical a feature of all scientific endeavor. The twin questions of what to measure and how to measure comprise a central problem of all scientific methods. Not only do these twin questions involve logical and semantic problems, but also mensural problems (i.e., what types of mathematical procedures to use) as well as technical problems of instrumentation. Furthermore, decisions regarding measurement significantly influence the type of analytic design that can be employed, as well as the cogency of the verification offered as proof of any test. For these and many other reasons, the heart of the scientific approach beats as strongly or as weakly as the sophistication of its measurements permits. In the chapters to follow all these various facets of the scientific approach will be related to the most critical aspect of the entire scientific enterprise, to wit: the test of hypothetical propositions by the use of an analytic design.

E. *Recording and Presentation*

VII.45   Since the whole purpose of scientific research is to add to man's store of verified knowledge, communication among scientists is a vital link in the whole chain of research effort. Science is now an international enterprise, and the communication of knowledge has become a highly complex process. Such communication is effected through personal correspondence, through published articles in scientific journals, monographs and books, and during meetings of the various professional societies. The nature of that communication, therefore, is a substantial part of the process of verification of one's findings. (This latter process will be discussed in the final chapter.)

VII.46   The proper recording of scientific data is influenced by the type of study: the quantity and quality of the material involved, the methods employed, and the interests and skills of the researcher. Though miniscule studies can be reported on a school boy's tablet, large-scale projects require huge electronic data-processing machines, whole buildings full of file cabinets, vast indexing systems, and a whole host of equipment and services necessary to a complete portrayal of the study. Furthermore, some studies are brief and discreet—e.g., an analysis of traffic patterns at a particular intersection, the verbal responses of a group of voters, the reaction of particular patients to particular drugs, or the expansion coefficient of particular metals. Other studies, however, may be quite continuous—e.g., Census studies, stock-market patterns, long-term analyses of basic organic processes, or the reactions of large groups of persons to new social conditions (i.e., prison, army life, resettlement, divorce, etc.). Certain general processes, however, are common to all types and sizes of studies, and certain general principles of effective communication apply to all mature scientific research.

VII.47   The arrangement and classification of data should follow the study design. Whether the study is basically descriptive or analytic, the data can be collected and arranged in a number of ways. Some workers prefer a cross-indexed card system, others prefer log books, while still others utilize work sheets. The only essential feature of any system is utility—i.e., the data should be arranged in such ways that they are clear and easily available. Competent researchers tend to develop neat and orderly systems of data arrangement, systematized in such ways that others would have no difficulty in studying the material if desired.

VII.48   Verbal data generally consist of written records or reports such as books, manuscripts, letters, diaries, documents, etc. Where utilized directly as primary sources, they should be available to other investigators either in libraries or in the possession of the researcher. Reference to original sources should clearly indicate, according to standardized methods of reference, the exact portions of the work being utilized. When written records are translated, the original must also be available to possible critics of the translation. When excerpts from a book, letter, diary or document are being utilized, the researcher must be careful not to interpret portions out of context—i.e., in such ways as to violate the intrinsic meaning of the whole. The data also should be collected according to some logical plan. Random collection of masses of facts in the hope that something unforeseen later might prove to be useful generally results only in confusion; and while being collected, the data should be arranged according to a well-thought-out system which permits clear and functional presentation. In cases where only one copy of an original source exists (e.g., diaries, letters), the cautious procedure suggests that the document be reproduced (e.g., by photostat or microfilming) as a safeguard against any possible loss of the original.

VII.49 Whenever feasible nonwritten verbal data should be recorded, preferably on tape; and even if the tapes have been transcribed, they should remain available to anyone skeptical of the transcription. Whether tapes or discs are used for recording, prudence sometimes suggests that they be duplicated even though they may later be transcribed. When editing of recordings seems desirable in order to eliminate irrelevant or extraneous material, it is especially imperative that copies of the original be kept somewhere on file for possible later perusal. In presentation, recorded data may be arranged either in transcribed form if only the content is pertinent, or *in toto* if inflections, timing, auxiliary sounds, etc., are deemed pertinent to the presentation.

VII.50 Visual data may be interpreted verbally, of course, especially when they are of a simple and indisputable type (e.g., the number of persons crossing a street, the size of dresses most frequently sold, or the color of packages most often selected). But when the quality of the data or their arrangement or interpretation are pertinent to their employment in the hypothesis, then the data should be recorded by appropriate instruments (cameras, photomicrography, macrophotography, X-ray, etc.). In some cases, sound movies (perhaps in color and perhaps even in stereo) are necessary; in other cases, color photographs or slides may prove desirable. Timed-sequence photographs, aerial views, stereoscopic slides, underwater photographs, telescopic photographs—the whole range of photographically recordable data should be exploited in a manner most pertinent to the needs of the study.

VII.51   In many instances visual data are transcribed into appropriate graphic presentations. Charts, graphs, maps, drawings, sketches, profiles, sociograms, etc., are often employed to permit visualization of concrete data. An elementary skill in methods of graphic presentation is essential to any competent researcher; and attractive as well as functional graphic presentations are a common feature of most well-done studies. In all cases where facts are presented graphically, however, the original data being depicted should be available in the files for possible inspection. Where the data are quantitative, the criterion of communicability should determine whether they might better be presented graphically or in tabular form; generally speaking, charts, graphs and maps are easier to comprehend than are tables.

VII.52   The particular advantages of charts and graphs over tables bear serious consideration. First of all, well-designed charts are generally much more effective in creating interest and visual appeal than are other types of presentations. Secondly, visual, spatial or relative data are more easily portrayed by charts or graphs than by any other method. Thirdly, charts and graphs permit an overall view of related data. Finally, charts and graphs can designate relationships probably better than can any other form of presentation. If there is any doubt about the advisability of depicting data in graphic rather than in tabular or textual form, the criterion of communicability should be employed by testing samples of the various forms of presentation to persons not familiar with the study at hand.

VII.53   Though quantitative data may be depicted in graphic form, in most instances tables are required if the data are relatively complex or detailed. Like the construction of graphs, charts, maps, etc., the construction of meaningful and clear tables is a basic skill required of all researchers who deal with quantitative data. Whether tables should be incorporated into the body of the text or grouped in an appendix will depend mainly upon whether or not they are absolutely necessary in order to understand the text. In most cases tables are relegated to an appendix, and only their interpretation is included in the body of the text.

VII.54 Physical data sometimes are basic to the study design, but are usually reproduced photographically. In some few cases, however, specimens, samples, prototypes, models, etc., may exhibit features lost in visual or mechanical reproduction, and in such cases the original data are appended to the study in the form of exhibits. These exhibits should be clearly indexed, neatly arranged and properly labeled so that any competent student of the subject can understand them and their relation to the study. Where original tools or instruments—e.g., scales, questionnaires, stimuli cards, measuring devices—have been devised and utilized, copies or samples of them also should be included in the exhibit. In determining which kinds and amounts of data should be included in a study, the scientist bears in mind the basic fact that scientific method demands exactness and clarity; and thus he includes in his presentation all those elements which a competent student of the subject might require in order to be able to understand and possibly criticize both the methods and the conclusions.

VII.55 Like any human being who might become emotionally involved in his field of interest, the scientist is at times apt to distort or exaggerate (by maximizing or by minimizing) his data in the direction of his predilections or prejudices. But the basic fact that scientific method is self-critical means that the possible distortions of an investigator can always be checked by the duplication or replication of a study by other investigators. For all practical purposes, then, no data or interpretations are acceptable as valid until corroborated by other investigators working independently. This means in effect that dishonesty cannot be practiced for any length of time; and therefore, that the presentation of a study should always make clear exactly how the study was done.

VII.56   When the study is ready to be written in its final form, adherence to a general set of established principles and practices enhances its clarity and hence its communicability. Though much scientific writing seems unnecessarily pedantic or obscure even to competent scientists functioning in the same field of interest, there is no legitimate excuse—other than sheer inability to communicate clearly—for much of the incomprehensible writing sometimes encountered in scientific journals. Effective writing in science combines skill in clear exposition with thorough yet succinct reporting of the essential features of the study. The scientist is concerned basically with the problem of communicating to his peers. He is not, as a textbook writer would be, concerned with writing for a group of novices, nor as a journalist or novelist would be, with attempting to reach the reading level of the layman. His style therefore optimally should combine the best features of technical reporting with the essential features of good composition.

VII.57   The preceding discussion suggests only a generalized outline of the principles underlying scientific reporting. It should be pointed out, however, that different fields—and different areas within various fields—operate on different levels of abstraction and conceptualization. In a field like history, for example, most discourse operates upon a relatively simple level of inference; but in an advanced field like physics or astronomy, the level of discourse has proceeded far beyond the common language of everyday use. In fact, the highly precise sciences now often discourse by means of *symbolic logic*—i.e., by the employment of mathematical methods which, with the use of special symbols of inference and denotation, enable the communicants to understand one another more precisely and more speedily than would be the case with ordinary language. Although symbolic logic has not yet become completely standardized in its symbolization, nevertheless the mathematical processes (usually calculus) underlying its procedures are universally understood by mathematicians.

Part III

SCIENTIFIC ANALYSIS

CHAPTER VIII

The Problem of Causation

A. ATTRIBUTING CAUSES

B. TYPES OF CAUSAL RELATIONSHIPS

C. LINKING CAUSATION WITH THEORY

A. Attributing Causes

VIII.1   The basic problem of science is the determination of causation. To primitive minds it may seem obvious to infer that A causes B simply because B always follows the appearance of A. But to the competent scientist, trained as he is in logical inference, it is not usually so clearly demonstrable that one factor actually *causes*—i.e., produces, or invariably results in, or is responsible for the appearance of—another simply because the second one happened to follow the first, or because the two are always found in each other's presence. For example, it is highly doubtful (at least to the trained scientist) that prayer causes rain simply because one prays consistently before the rain falls; or that the burial of amulets assures that a corpse will not rise from the grave simply because no corpse has ever arisen from its grave after amulets had been buried with it. Yet the consistency of such allied phenomena is sufficient "proof" to many laymen of a causal connection between such events. Such consistencies, whether individually or culturally experienced, tend to establish and reinforce popular notions of causality, and even to become crystallized into the folklore of a culture.

VIII.2   The notion of causation has various and highly significant logical, metaphysical and theological implications; but it is not pertinent to the purposes of this present discussion to examine these varied and many-faceted implications. To the scientist, the notion of causation is functionally meaningful only insofar as he is interested in ascertaining the kind (class or type), degree (amount), and direction (positive or negative) of relationships existing between two or more phenomena—in order to achieve his basic objective of predicting and possibly even controlling such phenomena. In other words, he is interested in ascertaining (a) which events induce—or are consistently followed by—other events (e.g., malnutrition results in disease); (b) the extent or frequency of such occurrences (e.g., in all cases, usually, sometimes, rarely); and (c) whether an event precedes or follows another, or perhaps even does both (e.g., truancy causes poor school performance, or poor school performance causes truancy). Such relationships, when ascertained to have a measurable degree of consistency, permit him to make reliable predictions and perhaps even to suggest changes deemed desirable in the event under consideration.

VIII.3   Causative statements, therefore, become practically synonymous with explanations of space-time events; for to explain something is essentially to give the reasons (i.e., to account for) why it occurs as it does. In physical sequential phenomena (i.e., those related by virtue of the fact that one or more follows others)—as in the case of clouds and rain, thunder and lightning, aging and death, etc.—an explanation is a statement of the causal connection between the events. For all practical purposes, then, the basic function of most analytic research designs is to explain in causal statements how various phenomena in question are interrelated.

VIII.4 There are several major types of *explanations* (i.e., causative statements or propositions). The ideal type (or "paradigm") may be termed—in logic—"formal" insofar as the fact to be explained (or "explicandum") is a *logically necessary consequence* of the propositions stated in the premises (of the total explanatory statement). In other words, given certain premises as either facts or assumptions, certain conclusions must necessarily follow according to the rules of deductive inference (see Chapter II). A simple example of this ideal type would be the explanation of why the men found in a lifeboat had died. The formal deductive argument in such a case might be stated somewhat as follows: A. Excessive deprivation of drinking water results in deadly thirst; B. the men in the lifeboat had been deprived of drinking water for an excessive period of time; C. therefore, the men in the lifeboat had died from thirst. Such a simplified explanation obviously could be elaborated in minute and factual detail in order to clarify all the significant terms and implications involved (e.g., How much deprivation is "excessive"? What is a "prolonged" period? What is or is not properly considered "drinking" water?). Furthermore, each of these specific elements of the argument could presumably be supported by empirical evidence relevant to human biological processes. The essential point, however, is that—given such propositions as major and minor premises—the conclusion must necessarily follow according to the accepted rules of deductive inference.

VIII.5 Another type of explanation is the "historical" or "genetic" type, which states that a given phenomenon is an inevitable consequence of certain specific antecedents. (It may not be known *why*, however, such a consequence is in fact "inevitable.") If, for example, it is established that eye color is a biologically transmitted trait, then it follows that a child's eye color is "caused" by the eye color of his parents. Likewise, if it is contended that advances in technology always have been followed by increasing industrialization (and, hence, by urbanization), then it follows that increasing technical invention will "cause" an increase in industrial employment (and, hence, in urban growth). It should be noted that this type of explanation differs from the preceding type insofar as the consequence in the second type does not *necessarily* have to follow the antecedent conditions or events. (It may be that the sequential linkage of events is only historically coincidental.) Nevertheless, this type of explanation is often employed in the absence of a logically superior (i.e., more cogent, hence more convincing) type, particularly when there seems to be some reasonable connection between the antecedents and the consequences.

VIII.6 A still different type of explanation is termed "extrapolative" or, more often, "probabilistic." In this case an event is "explained" on simple empirical grounds without resort to any implications of inherent or necessary causal connection. In other words, the premises in such an explanatory statement do not formally imply the consequence; but the premises do contain certain statistical assumptions. The *explicandum*, then, is an individual instance of one of the assumptions. Thus, for example, it may be found that higher-income families have more telephones *per capita* than do lower income families; that urban dwellers have more telephones *per capita* than do rural dwellers; or that "white-collar" workers have more telephones *per capita* than do "blue-collar" workers. Given such facts it would be simple to "explain" why—and therefore to predict that—the *per capita* number of telephones will increase when it is found that there are now proportionately more upper-income families, or more urban families, or more "white-collar" families.

VIII.7 In such a type of explanation, the consequence is "explained" simply by virtue of the fact that an "If this, then that" connection has been found. Though obviously of great practical value for purposes of prediction, such explanations are inherently weak due to the absence of any established relation in terms of "necessary" or "sufficient" conditions related to the "cause." In short, not knowing "why" such connections between the antecedent and the consequence exist, there is no assurance that such connections *must* or *always will* exist; and therefore that such trends will necessarily continue. Further consideration of this type of causal connection will be deferred until the discussion of probability correlations later in the chapter.

VIII.8   A distinctly different type of explanation from those discussed above is commonly designated as "functional" or "teleological." A confusion often occurs in this case because these two terms are not necessarily synonymous. Strictly speaking, a *functional* explanation is one in which the argument implies that certain consequences are a result of certain functions or processes inherent in a given system. Thus, for example, the explanation of why an engine is equipped with an automatic speed regulator ("governor") might be stated in functional terms somewhat as follows: "because without a governor the engine might be run too fast, and hence break from excessive vibration or centrifugal force." It should be noted in this case that the "why" of the explanation is related to, and only to, a condition inherent in this particular system (viz., of engine breakdown in relation to speed). An extended example of a functional explanation would be the case of the eraser on the end of a pencil. It could be "explained" by: (a) assuming that some people make mistakes in writing with a pencil; (b) assuming that it is desirable to remove or erase such mistakes; (c) granting that a piece of rubber will erase pencil mistakes; (d) assuming that it would be convenient to fasten a piece of rubber on the end of a pencil in order to erase mistakes; and (e) deduce that that is why erasers are in fact placed at the top of pencils.

VIII.9   A *teleological* explanation, on the other hand, although it may also serve this functional purpose, implies certain ends or means-to-an-end to be served in a system of causal connection. In some extreme cases of teleological explanations it is not uncommon to explain a consequence by implying that it occurs in terms of "original" or "final" causes somehow related to the intentions of a divine, supernatural or extra-natural agent. A simple example of such an explanation might be the answer to the hoary question: "Why does the sun shine?" "Because God makes it shine." "Why does He make it shine?" "In order to give us light and warmth." "Why do we need light and warmth?" "In order to live." "Why do we need to live?" "Because God wants us to." The essentially teleological form of this type of explanation is found in the final statement; a statement that something occurs in order to serve an intended purpose.

VIII.10 A clearer distinction between a truly functional and a teleological explanation can be seen in an example from biology. Suppose, by way of illustration, that the question is asked: Why do humans have hair on the top of their heads? The answer in functional terms might be something like: In order to provide insulation for the head against excessive sunshine or cold, thus minimizing sunstroke or freezing. Such an answer, it should be noted, parallels the previous example of the engine and its governor, in that in both cases the explanation is related to understanding the functioning of a whole system (i.e., the running engine in one case and the living human in the other). In either case such an answer implies a statement of fact: in the first case that the governor does restrain the engine from achieving excessive speed, and in the second case that head hair does minimize the danger of sunstroke or freezing. (Such contentions could, as statements of fact, be tested empirically by the analytic method to be examined later in this chapter.) But such an answer (in the case of head hair) could also carry a teleological implication if it *assumes* that somehow, or by some agent or agency (usually God or "nature"), it is desirable or inevitable that man's survival be promoted or guaranteed.

VIII.11 Such an assumption is not, of course, an empirical proposition; and therefore cannot be tested for factual validity. It is simply inferred—or imposed by virtue of one's belief—that there exists a natural or Divine purpose in the world (and, hence, in man's existence). And while a truly functional explanation may suffice in many sciences, a truly teleological or "purposive" explanation of the type illustrated here is rejected in the more advanced sciences. The reason for this rejection is twofold: in the first place, a teleological explanation does not conform to the ideal (deductive) model examined earlier, for it imposes assumptions (about causative agents) that cannot be tested empirically, and thus prohibits the explanation from ever being supported on an objective or factual foundation. In the second place, and even more important, such an explanation is unparsimonious— i.e., it adds nothing which is either useful or necessary to the explanatory statement.

VIII.12 In our example of the engine, it was necessary to explain the function of the governor in order to understand why the engine did not speed itself into demolition; but in the example of head hair, no such purpose was served by the explanation. To argue that head hair is necessary for, or contributes to, human survival—besides being questionable in fact—infers that human survival can be understood (i.e., explained) by such statements of contributory causes. Actually, however, the role of head hair can be adequately designated by a simple factual statement: viz., it either does or does not contribute to human survival. To go beyond this factual assertion is to include metaphysical connotations (about "purpose") which open the door to unlimited speculation about Divine, natural, or "original" causes of man's existence; and such speculation, however interesting in itself, contributes nothing useful to the body of an empirical science. In short, an explanation serves a useful and even necessary purpose in science when it helps to establish a sequence of causal connections between events; but it serves no useful purpose if it merely, or even additionally, introduces speculative connotations of a metaphysical sort. Such connotations, however interesting to a philosopher, contribute nothing useful to the solution of the scientist's problem of ascertaining causal connections in order to make reliable predictions of sequential events. And for this reason alone, teleological explanations are avoided in all the advanced sciences, however they may be employed to lend seeming plausibility to the more immature disciplines.

B. *Types of Causal Relationships*

VIII.13 The scientist prefers to approach the problem of causation by ascertaining (a) the type of relation existing between and among phenomena, (b) the necessary and/or sufficient conditions related to particular phenomena, and (c) the degree of interrelation found between two or more phenomena in terms of probable occurrence. In terms of (a), the type of relationship found existing between two or more phenomena can be usefully designated in terms of five subtypes as follows: The first type indicates that A is the cause of B—i.e., whenever A is found, B always follows. (For example: A, interest in grades, causes B, studying; see Figure 2, a.)

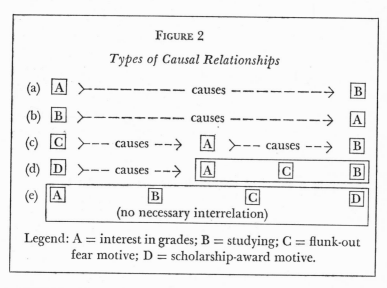

FIGURE 2

Types of Causal Relationships

(a) [A] >——————— causes ——————→ [B]

(b) [B] >——————— causes ——————→ [A]

(c) [C] >—— causes ——→ [A] >—— causes ——→ [B]

(d) [D] >—— causes ——→ [A]   [C]   [B]

(e) [A]   [B]   [C]   [D]
(no necessary interrelation)

Legend: A = interest in grades; B = studying; C = flunk-out fear motive; D = scholarship-award motive.

VIII.14 The second type of relationship is the converse of the first, namely, that B is the cause of A. (For example: B, studying, causes A, interest in grades; see Figure 2, b.) The third type of relationship indicates that both A and B are caused by C. (For example: A flunk-out-fear motive, C, causes an interest in grades, A, which in turn causes studying, B; see Figure 2, c.) The fourth type of relationship indicates that A or B are found only in the presence of C, and therefore are *incidentally* related by this fact, but are caused by D. (For example: Interest in grades, A, and studying, B, are found only when a flunk-out-fear motive,

C, is present; but the essential cause is D, a scholarship-award motive; see Figure 2, d.) The fifth type indicates that A and B, or A, B and C, or A, B, C and D are found together only by *coincidence* or "chance." (See Figure 2, e.) Actually, other combinations of causal relationships are possible; but only these— since they are basic—need concern us at this time.

VIII.15 Ascertaining the necessary and/or sufficient conditions which induce a phenomenon to occur relates the type of causative relationship (discussed above) to the *possibility* of its occurrence. A factor may be regarded as causally related by *necessity* to another if the second one cannot take place without the first. (Water, for example, is necessary before one can cause plants to grow.) But though a factor may be necessary for the causation of another, its presence alone may not be *sufficient* to induce the other to occur. (The mere presence and availability of water, for example, gives no assurance in itself that it will be employed for the growing of a plant.) In short, a factor is regarded as a necessary condition for the causation of another if the first one must be present for the second one to occur; or, conversely, if the second one cannot occur without the first one. But a factor is regarded as a sufficient condition for the causation of another if the second one always follows—or is always found in the presence of—the first one. Therefore, the scientist is fundamentally interested in ascertaining both the necessary and sufficient conditions related to the phenomena which he is studying as dictated by the demands of his hypothesis.

VIII.16 Ascertaining the *probable occurrence* of a phenomenon avoids all the implications of causality heretofore discussed; and simply attempts instead to ascertain the degree of probability that one or more factors will be found consistently together. The statement, for example, that "The chances are nine out of ten that keeping the children after school will cause a significant increase in discipline problems," expresses a demonstrably verifiable fact of relationship. Functionally, such a statement serves the scientist's basic purpose of ascertaining the kinds and degrees of relationships existing between selected phenomena—provided, of course, that it is combined with a designation of the type and/or direction of relationship, as discussed previously, and with a determination of the necessary and/or sufficient conditions involved.

VIII.17 A statement of probable occurrence is based upon either an *a priori* theory or a *relative frequency* theory. The first type is so called because the knowledge required for the expression of the probability of occurrence can be (and in fact must be) achieved *before* (i.e., *a priori*) the event. Suppose, for example, that it is known that a door prize will be awarded to one person in an audience of five hundred. Assuming that everyone has an equal chance to win the prize (i.e., that the chances are "equiprobable"), the probability that any particular person will win the prize therefore can be deduced simply as one chance in five hundred. Conversely, the probability that any given person will not win the door prize would be four-hundred ninety-nine to the one that he would.

VIII.18 The second type of theory also expressed a relationship as probable, but on the basis of experience derived from observing the frequency of occurrence—not on the basis of *a priori* calculation. Thus an observer might note that every fourth man entering a hall is wearing a bow tie. The probability that the next man who enters will be wearing a bow tie therefore could be expressed as one chance in four. There is no assurance, however, that all of the following men entering the hall will be wearing bow ties in the ratio of one in four. The inference made is probable only to the degree that the sample observed is valid (in the various ways as discussed in Chapter V). Since no event has any inherent or intrinsic probability of its own, the frequency of its occurrence can be determined only in terms of the evidence upon which it is based as both relevant and relatively sufficient.

VIII.19 To summarize this discussion of causation, it might be said that causative statements serve two major purposes in science: (1) to link sequential events in predictable patterns, and (2) to suggest interrelationships between and among various phenomena. Whether the "explanation" involves a causal proposition or simply a statistical probability statement, it serves the analytic function of science by permitting the investigator to go beyond simple and mere description to the discovery and establishment of reliable relationships.

C. *Linking Causation With Theory*

VIII.20　At this stage of the total scientific enterprise, several avenues to the analytic approach are possible in order to link the discovery of reliable relationships with the theory or hypothesis motivating the research. First, the investigator may start with a verified fact (e.g., the volume occupied by a gas varies inversely with the pressure exerted upon it), and then proceed to relate it meaningfully to other verified facts (e.g., air pressure varies with temperature, or atmospheric pressure varies with altitude). At this stage of knowledge a science has not advanced very far from the kind of reliable knowledge possessed by many primitive tribes; although obviously such rudimentary knowledge is the foundation of all the advanced sciences. In many of the rudimentary sciences of today (ranging from seismology to anthropology), the level of development has not advanced significantly beyond this point.

VIII.21　Second, the investigator may start with a hypothesis (e.g., the majority of women evidence less interest in mechanical topics than do men) and then proceed to test it in order to confirm it in terms of verified facts. (For example, vocabulary tests might show that women are less familiar with mechanical terms than are men; or that women, on the average, do not learn to repair cars as quickly or as effectively as do men.) This stage of development characterizes many of the sciences today, particularly the behavioral sciences, but also some fields of medicine, genetics, archeology or paleontology. Admittedly of a higher order—i.e., more complex and usually more detailed—than the preliminary level of development mentioned previously, a science at this stage requires the synthesizing power of more general and inclusive notions of interdependence or interrelationship ("theories") before it can reach the mature stage of the advanced sciences.

FIGURE 3

Method of Agreement

(a) Positive form

Factors in situation								*Phenomenon*
(1) A	B	C	D	*E*	(Etc.) ⟵--- result in ---⟶ x
(2) *E*	F	G	H	I	"	"	"	"
(3) J	*E*	K	L	M	"	"	"	"
(4) N	O	*E*	P	Q	"	"	"	"
(5) R	S	T	*E*	U	"	"	"	"

(b) Negative form

(1) A	B	C	D	*E*	" ⟵——— "	"	⟶ *no* x
(2) *E*	F	G	H	I	"	"	"	"
(3) J	*E*	K	L	M	"	"	"	"
(4) N	O	*E*	P	Q	"	"	"	"
(5) R	S	T	*E*	U	"	"	"	"

(c) Combined form

(1) A	B	C	D	*E*	" ⟵——— "	"	——⟶ x
(2) A	B	C	D	*E*	"	"	"	*no* x
(3) R	S	T	*E*	U	"	"	"	x
(4) R	S	T	*E*	U	"	"	"	*no* x

Legend: A = time of day; B = season; C = geographical lo-
cation; D = sex of the raincoat wearer; *E* = wearing of a raincoat; x = rain; *no* x = no rain; F =
age of the wearer; G = educational level of the
wearer; H = religion of the wearer; J = height of
the wearer, etc.

VIII.35   In many instances of inductive inference it is just
as desirable to ascertain what factor may be related to the *absence*
of a phenomenon rather than to its presence; or, in other words,
it may be desirable to ascertain what factors *prevent* a given
phenomenon. In such cases the method of agreement may be
stated in its negative form as follows: If two or more cases of the

absence of a phenomenon have *only one* factor in common, then that factor may be regarded as the cause or as the effect of the absence of the phenomenon. For example, if all people who do *not* have dental cavities have fluoride in the drinking water, then it reasonably may be inferred that there is a causal relation between the absence of cavities (the phenomenon) and the presence of fluoride in the drinking water (see Figure 3, b).

VIII.36 A basic feature of the inferences made in the above diagrammatic examples should be carefully noted—for this feature is inherent in all analytic designs. This feature is the *undirectionality of causal connection*. In the diagrams above, the only defensible logical inference that could be made from the associations noted is that one factor (*E*) either is or is not causally related to another (x) or (*no* x); the connections (viz., between *E* and x or between *E* and *no* x) do not indicate the *direction* of causality—i.e., they do not indicate which item (*E* or x) is "cause" or "effect." The only way to answer the question of "cause" *versus* "effect" is in terms of sequential appearance—i.e., which of the two factors *must precede* the other as either a necessary or sufficient condition.

VIII.37 The method of agreement is basic to the process of inductive inference; but if used alone is severely limited for the purpose of inferring causal relationships. In the first place, it requires instances which are *unlike* in every respect save one; and such instances are often very difficult either to obtain or to ascertain. In the second place, this type of proof (of causal connection) can only *suggest* that the hypothesis being deduced is *probable*, but *not certain*. To achieve a more significant degree of certainty other methods of inference must be employed. In the third place, this method can be generally applied only in those instances where a *single* cause is involved; it is not applicable to problems involving multiple causation.

VIII.38 Some of the shortcomings of the method of agreement are overcome by the *method of difference*. This method states that if two sets of circumstances or situations *differ* in *only one* factor,—and the one containing that factor results in a particular consequence while the other does not—then that one factor can be assumed to be the cause of the consequence. For example, if two cars are identical except that one has an overdrive while the other does not, and if the one with the overdrive gives better mileage than the other (when driven under identical conditions, of course), then it may be assumed that the single factor of difference (i.e., the overdrive) may be regarded as the cause of the consequence (viz., better mileage, see Figure 4).

FIGURE 4

Method of Difference

Factors in situation   *Phenomenon*

(1) A B C D E   (etc.) ⟩———— result in ———→ x
(2) A B C D *no* E   (") ⟩———————   "   "   ——→ *no* x

Legend: A = make of car; B = color of car; C = sex of the
    driver; D = time of day; E = overdrive; etc., =
    various road conditions; x = better mileage.

VIII.39   Two essential limitations of the method of differences should be noted: (a) It requires instances which are *alike* in every respect save one. This is the converse of the limitations of the method of agreement. (b) More significant, however, is the ever-present possibility of committing one of the most serious yet common of all errors of logical inference: namely, the *post hoc* (*ergo propter hoc*) fallacy. Literally "after this, therefore because of this," the *post hoc* fallacy most often is observed in the case of primitive reasoning which argues naively in a direct cause-to-effect sequence. (For example: I broke a mirror, then I had bad luck; therefore breaking the mirror was the cause of bad luck.)

VIII.40   To minimize the possibility of committing the *post hoc* fallacy, the method of difference can be employed in its converse form, and as such may be stated as follows: A factor *cannot* be the sole cause of a consequence if the consequence *does not always* take place when the supposed cause does. In other words, if a given consequence (e.g., bad luck) does not always follow a presumably causal antecedent (e.g., the breaking of a mirror), then the consequence cannot be said to be caused by the antecedent event. (We are here concerned with direct causation as defined earlier and illustrated in Figure 2, a. We are not considering incidentally or coincidentally allied causes.)

VIII.41 The foregoing discussion has attempted to present in simple form the inductive inferences designated as the method of agreement or the method of difference. It should always be borne in mind, however, that such inferences are relatively primitive, hence not very useful, unless employed in an analytic framework which designates the necessary or sufficient conditions involved in the statement of causality. In the above example of the broken mirror and bad luck, even the converse form of the method of difference—employed here to minimize the *post hoc* fallacy—ignored the question of necessary or sufficient conditions. It is conceivable, for example, that the mirror did not break because the subject was alert to the possibility of slipping. If being so alert is a sufficient condition to prevent ("non-cause") the occurrence of bad luck—or, conversely, if the absence of alertness is a sufficient condition to insure the presence of ("cause") bad luck—then the method employed (either agreement or difference) must include such a factor within its analytic framework. To avoid it is to sterilize the design of any possibility of achieving inferential fruitfulness.

VIII.42 The various forms of the methods of agreement or difference just presented—or the combination of them in the so-called "joint" method—exhibit two other major shortcomings independently of those mentioned. First, it is extremely difficult, and in some cases impossible, to exclude or to isolate possible causes from related factors or from constantly changing factors. It is impossible, for example, to isolate the tides, or daylight, or aging, or death—in order to subject them alone to deductive inference—by comparing them with instances where tides, daylight, aging or death do *not* occur. Secondly, it may be impossible to isolate interrelated factors by separating them from each other in order to study each separately. For example, it is impossible to separate the tides from the influence of the moon, or daylight from temperature changes, or aging from metabolic processes, or death from coronary conditions. Furthermore, it is noteworthy that all these methods are *qualitative*—i.e., they suggest probable causes *only when* factors can be isolated. However, if probable causes are suspected to exist, then a *quantitative* method may be helpful to increase the probability that a causal relationship exists—even if the causes cannot be isolated.

VIII.43 Such a method is called the *method of concomitant variation,* and can be stated as follows: If, in two or more cases of related phenomena, a *change* in either the *direction* or *amount* of a factor is *consistently accompanied* or *followed* by a change in another factor, then the two factors are causally related—i.e., one is either the cause or the effect of the other. For example, if an increase or decrease in the amount of food eaten is followed by a *concomitant* increase or decrease in one's body weight, then it may be assumed that the amount of food eaten is causally related to body weight (see Figure 5).

VIII.44 However, an increase in body weight (say, for example, from hardening of the muscles due to exercise) may be the actual cause of the increased amount of food eaten, not the other way around. This method, therefore, like the others mentioned,

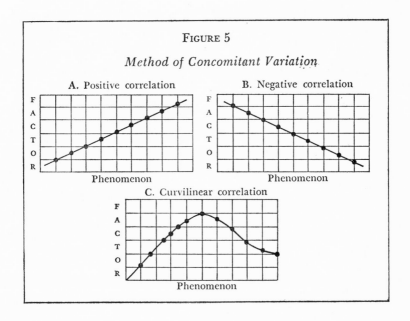

FIGURE 5

Method of Concomitant Variation

A. Positive correlation   B. Negative correlation

C. Curvilinear correlation

does not indicate which factor is the cause of the other; but only that the two factors are somehow interrelated beyond coincidence or chance. Since this method can be employed only when quantities or magnitudes of effects and causes can be distinguished, it demands some objective techniques of measurement and usually also requires the employment of a statistical framework of presentation and analysis.

VIII.45 The final type of inductive inference, the *method of residues*, more specifically resembles a deductive inference than do the other analytic methods just discussed. Essentially, the method of residues can be stated thus: When the component factors of a situation are known to result in a given consequence, and when the influence of all but one of the specific components is known, then the influence of the remaining factor can be deduced from the total consequence. For example: It has been established that a car weighs 3,000 lbs. A scale is placed under each of the three wheels, and the combined weight is 2,100 lbs. The weight upon the fourth wheel, therefore, can be deduced to be 900 lbs. (This could occur in the case of an unbalanced load.) Schematically, the method of residues can be represented as follows (see Figure 6).

VIII.46 The method of residues differs in two essential respects from the methods of agreement, difference, or concomitant variation. First, it may be employed in single cases of a phenomenon or consequence, while the others require two or more cases for comparison. Second, it demands previous knowledge of the consequence, and in this respect is simply a form of elimination.

FIGURE 6

Method of Residues

Factors in situation		Consequence
A + B + C + D $\longleftarrow$ --- results in ---$\longrightarrow$		wxyz (3,000 lbs.)
A + B + C $\longleftarrow$ —— " " ——$\longrightarrow$		wxy (2,100 ")
Inference: D $\longleftarrow$ —— " " ——$\longrightarrow$		z (900 ")

In many instances, however, it can be a very useful tool of analysis. Let us suppose that one is interested in determining the specific cause of vibration in a piece of machinery. Let us also assume that all the possible causes are known beforehand (e.g., loose mounting bolts, irregular mounting platform, unbalanced shaft, loose windings, vibration in the building). Then it becomes a simple matter to test each separate factor independently until the specific causal one is found. It might be, of course, that there is not just one, but two or more possible or interrelated factors involved in causing the consequence. In such a case the method of residues would simply require a more complex framework of analysis—viz., testing combinations of factors—but essentially would remain a problem in subtraction.

VIII.47 A word of caution about the employment of these inductive methods might bear emphasis at this point. The determination of causal relationship is the heart of any science. But a servile faith in the efficacy of these methods to discover causal relationships exhibits a naive understanding of their function in the total pattern of inference about phenomena. Since the selection of possible factors of relatedness is almost infinite, the search for highly probable causes can result in inordinate and often fruitless expenditures of time and effort. (The story of Dr. Erlich's discovery of a specific inhibitor for the spirochete of syphilis is a case in point. He purportedly tried six hundred and five compounds before hitting upon salvorsan, or "606." The best example of simple "trial and error," however, is the centuries-old yet fruitless search of the alchemists for the "elixir of life.") The selection not only of possible but also of probable causal factors is a function of the theoretical and hypothetical framework of the research design—not of the analytic *test* of the hypothesis. In short, these methods are useful only to the degree that the hypothesis provides them with relevant factors of both possible and probable causal connection.

CHAPTER IX

The Analytic Method

A. THE STRUCTURE OF ANALYTIC DESIGNS

1. *The basic question of analysis*
2. *Analysis basically simple*
3. *Laboratory verification*
4. *The basic analytic design*
5. *Variations of this basic design*
6. *The before-after design*
7. *The* ex post facto *design*
8. *Usefulness of this design*
9. *The comparative method*
10. *Ambiguity of this designation*
11. *Abuses of the comparative method*
12. *Further abuses of this method*
13. *Distortions of this method*

B. THE ROLE OF VARIABLES

14. *The basic problem*
15. *Variables may be unsuspected*
16. *Controlling variables*
17. *Controlling relevant variables*
18. *Matching variables*
19. *Matching by frequency distribution*
20. *Randomization*
21. *Random sampling vs randomization*

C. QUANTIFICATION

22. *The basic problem*
23. *Statistical skill needed*

204

A. The Structure of Analytic Designs

IX.1 The fundamental design of analytic method employs the various methods of inductive inference discussed in the previous chapter. This design is basic to all scientific analysis, regardless of the specific techniques employed (e.g., experimental, comparative, historical). Essentially, the question of analysis is a matter of determining the *degree* and/or the *type* and/or the *direction* of relationship existing between two or more phenomena under given conditions. The problem to be solved, then, is arranged according to various techniques all calculated to utilize this basic design. The analytic design achieves its best known form in the controlled experiment; and it may be for this reason that the so-called experimental method is often thought to be the only legitimate form of inductive inference. It should be noted in the discussion to follow, however, that although the controlled experiment demonstrates the optimal form of inductive inference, the analytic design can be (and often is) achieved outside the laboratory, and without all the usual refinements of the typical physical experiment.

IX.2   In its simplest form, all that is required in the analytic method is that the researcher make a series of observations of two apparently related factors. According to the methods of inference discussed, he can then determine if a relationship exists by noting whether one factor changes, or appears or disappears, consistently in the presence or absence of the other. Every person employs this method of establishing relationships in his everyday life. For example, he notices that the flick of a wall switch is immediately and consistently followed by the glow of the light bulb; that the turn of a handle immediately and consistently results in a flow of water from a faucet; that winding his broken watch does not cause it to begin running again; or that kicking the front tire of his stalled car does not cause the dead engine to start. (These are all primitive forms of the method of agreement or difference.)

IX.3   When the methods of difference are employed in a laboratory, however, verification is aided by the employment of a "control" group tested against (i.e., compared with) a "variable" or "experimental" or "test" group. This means, in effect, that one sample is held constant (i.e., controlled to prevent possible change) while a comparable sample is subjected to the influence of the presumed causal factors being tested. According to the various principles of inference discussed, it may therefore be reasonably assumed that any changes which occur in the test group—but which do not correspondingly occur in the control group—must be consequence of the causal factor; for the consequence or effect always occurs when it is present, but never occurs when it is absent. In a classical laboratory demonstration of this method, for example, two comparable feathers are allowed to drop a given distance within two tubes of identical size. Both feathers are found to fall at the same rate. Then the air is removed from one tube (the "variable") but not from the other (the "control"), and the two feathers are dropped again. This time the feather in the vacuum falls faster than the other; thus indicating that the air (which creates resistance to the feather) is the "cause" of the slower rate of fall. Other methods differ from this basic one only in the respect that they may not actually "control" either element of the design—i.e., they may instead measure the differences existing previous to the introduction of the influencing factor, and then note the kind and degree of changes which occur after the presumed causative factor has been applied. Such methods, however, depend upon the same principles of logical inference and employ the same principles of inductive analysis as does this basic design.

IX.4 The demonstration of causal connection by means of the analytic design rests upon a group of essential premises: namely, that two matched samples of a phenomenon are available; that one will be controlled while the other is systematically varied by being subjected to the influence of a stimulus as posited by the hypothesis; and that changes occurring in the experimental group can thereby be attributed to the influence of the stimulus (see Figure 7).

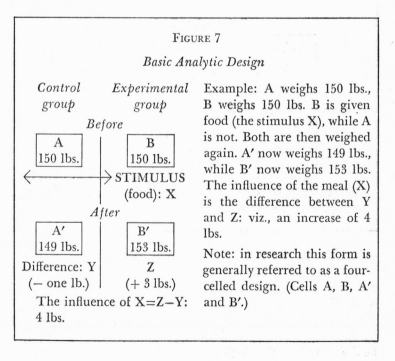

FIGURE 7

Basic Analytic Design

Control group	Experimental group
Before	
A 150 lbs.	B 150 lbs.
←	→ STIMULUS (food): X
After	
A' 149 lbs.	B' 153 lbs.
Difference: Y (− one lb.)	Z (+ 3 lbs.)

The influence of X=Z−Y: 4 lbs.

Example: A weighs 150 lbs., B weighs 150 lbs. B is given food (the stimulus X), while A is not. Both are then weighed again. A' now weighs 149 lbs., while B' now weighs 153 lbs. The influence of the meal (X) is the difference between Y and Z: viz., an increase of 4 lbs.

Note: in research this form is generally referred to as a four-celled design. (Cells A, B, A' and B'.)

IX.5 In many instances, however, this logically optimal procedure is impossible—or too difficult—to effect, particularly when matched samples cannot be obtained, or when time changes the constants demanded by the experimental design. In such cases variations of this "four-celled" design (one "cell" each for both the experimental and the control group before and after the stimulus) are employed; and to the extent that they differ qualitatively from it, they can be regarded as of lower precision in relation to the optimal procedure.

IX.6 The most common variation of the basic four-celled analytic design is the so-called "before-after" design. In this case a sample is measured, then subject to the influence of the stimulus, then remeasured (see Figure 8). The weakness of this pro-

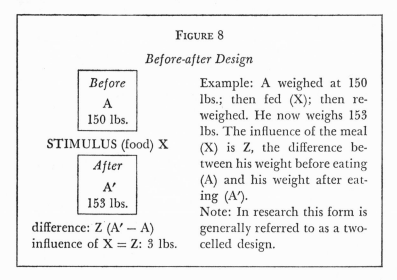

FIGURE 8

Before-after Design

Before
A
150 lbs.

STIMULUS (food) X

After
A'
153 lbs.

difference: Z (A' − A)
influence of X = Z: 3 lbs.

Example: A weighed at 150 lbs.; then fed (X); then reweighed. He now weighs 153 lbs. The influence of the meal (X) is Z, the difference between his weight before eating (A) and his weight after eating (A').

Note: In research this form is generally referred to as a two-celled design.

cedure lies in the fact that it is quite possible that whatever changes may occur might also have occurred if the stimulus had not been introduced. In other words, the changes might have occurred in the intervening time due to factors unsuspected and therefore uncontrolled and unaccounted for. A common and yet meaningful example of this type of inferential error might be the case of "explaining" the influences of a college education.

Suppose, for example, that a group of entering college freshmen are measured on a scale of political knowledge; and then measured again on the same scale after graduation four years later. An increase in their average score might suggest to the unwary that four years of college education had "caused" them to improve in political knowledge. Actually, of course, the reverse might be true. That is, the increase in average score might have occurred anyway due to other factors taking place during those four years. In fact, a control group (i.e., comparable, but not going through college), might actually have shown an even higher average score after the same four years (due, possibly, to maturation, or to a public increase in political awareness). The employment of a control group is the only way to solve this possible and always logically present dilemma.

IX.7 Another common variation of the basic four-celled analytic design measures only the "after" group and then proceeds to determine as well as possible the presence or absence of comparable relevant factors found in the hypothetical "before" group. This so-called *ex post facto* (i.e., after the fact) design is utilized mainly in those instances where an event already has occurred and cannot be repeated or replicated. In order to determine either or both the qualitative or quantitative influences of the event (e.g., the effect of an earthquake upon houses), it is necessary to match the affected group with a comparable group which existed before the event occurred, just as is done in the before-after design. The weakness of this method, of course, lies in the fact that comparability cannot be verified, since *two matched groups* both existing before the event occurred were *not* obtained. The houses that fell during the earthquake, for example, may have appeared to have been comparable to those standing before the earthquake hit; but one can never be certain that other unknown variables were not involved. Termites, for example, may have weakened the fallen houses so that a gust of wind at the time, and not the earthquake, could have caused the damage.

IX.8 In those cases where "before" comparability can be ascertained, however, this method is just as valid and as useful as is the basic four-celled analytic design. The *ex post facto* method has proved its usefulness in many instances where no other method could have been employed. In the biological sciences, for example, changes in average weight, height, athletic performance, morbidity, mortality, fertility, etc., have been studied fruitfully by comparing present populations with presumably comparable populations that lived in previous times. In the social sciences, changes in crime rates, in marriage and divorce rates, in schooling, in traffic accidents, in alcoholism, in drug addiction, etc., also have been studied profitably by this type of design. This method certainly is one of the most commonly employed in many fields of actuarial statistics, where such comparisons have been made of the relative earning power of college graduates with non-college graduates (who presumably were comparable before one group went to college), of World War II veterans with World War I veterans in terms of eventual employment, and of delinquency tendencies in terms of changing environmental influences.

IX.9   Although the essential process of analytic method is a comparison of two or more groups in relation to an intervening variable, in practice the term "comparative" method is commonly employed to denote the technique of comparing two or more groups which do not permit of manipulation. Actually, the so-called comparative method is a version of the basic four-celled design. In its bare essentials, the *comparative method* studies two groups presumably matched in all relevant and common attributes except one (the "variable"). Then the cause (i.e., the influence producing the difference in question) is inferred. In a classical example from social psychology, the problem of adolescent emotional disturbances was studied by this method. It was noted that many American children found adolescence a period of great emotional stress; whereas in many primitive cultures adolescence was not accompanied by like consequences. Here culture is the variable. Since the physiology of adolescence is presumably common to all peoples—regardless of such factors as race, diet, nationality, climate, geography, religion, etc.—it was deduced that the cause of the disturbances must be sought in cultural and not in physiological differences.

IX.10   The neophyte scientist should bear in mind that the "comparative method" does not actually *compare* in the precise sense that the basic four-celled design does. The comparative method does not permit manipulation of the intervening influence, and it does not maintain a true control group. Furthermore, it does not permit replicability—one of the most basic of all tests of verification. Whether employed directly—as in the comparative studies in history, ethnology, sociology, geology, biology, ecology, etc.—or indirectly—as in the projective techniques employed so dramatically in psychology and psychiatry—validation of causal inferences by employment of the comparative method demands the same logical and factual conditions as does the basic four-celled design.

IX.11   Abuses of the comparative method are all too common. Two examples might serve to illustrate such abuses in reasoning, particularly when this method is employed by laymen. The first example can be found in everyday use by persons who commit what is called the "animistic fallacy"—i.e., the fallacy of ascribing human feelings to inanimate substances (e.g., "the sighing wind," "the angry mood of the sea," "the laughter of the brook," "the tears of the early morning dew"). Though such phrases are often simple literary devices of a metaphorical sort, to many persons who knowingly or unknowingly believe in *animism*, such expressions connote an illogical comparison of unlike phenomena.

IX.12   A more significant abuse of the comparative method is *anthropomorphism*—i.e., the practice of attributing specifically human qualities to the lower animals, (e.g., attributing human "love" to the protective practices of a mother bear; referring to a hyena as "laughing"; speaking of the "domesticity attitudes" of some birds; or implying that one dove "mourns" the absence of its mate). Allied with this practice is the comparable practice of *zoomorphism*—i.e., the error of attributing specific lower-animal qualities to man (e.g., referring to a man's "animal passion", to a woman's "instinctive protectiveness" toward her offspring, to a boy's "natural" fear of fire, or to a woman's "mating instinct").

IX.13   Such practices are a distortion of the comparative method; for though man and the lower animals are comparable in some demonstrable aspects (as witness the fruitful experiments made on animals which have applicability to human situations), they are just as often incomparable. Thus to argue that such behaviors as maternal protection of the young, mating tendencies, nest-building instincts, gregariousness, "pecking order," aggressiveness, homing instincts, spawning instincts, hetero- or homosexual pairing, etc. (which appear to be innate in many lower animals), therefore must be "natural" also to man certainly pushes the comparative method in such cases to very illogical ends. Yet the history of science, particularly of the behavioral sciences (and especially of psychology), is rife with errors committed in the name of the comparative method. In the chapter to follow the verification criteria of analytic method will suggest why such errors as those illustrated are so often made in the name of science.

B. *The Role of Variables*

IX.14   Basically simple, the analytic methods at times pose some highly vexing problems. The first is that of both recognizing and controlling all the relevant variables involved in the hypothesis. Recognition may be frustrated simply because the variables are difficult (or at times actually impossible) to find; and this is particularly true when they are only incidentally related to the phenomenon (refer to Figure 2, e in the previous chapter). Consider, for example, some of the well known cases from the history of science. In the case of malaria, discovery both of the general cause (viz., a bite from a mosquito) and of the specific cause (viz., a bite from an infected female of the species *Anopheles*), required years of concerted effort by teams of scientists from many nations. The arduousness of the search was due to the fact that the mosquito was not suspected as a cause in the first place, and secondly because not all mosquito bites resulted in malaria. In the case of certain bodily malfunctions, discovery of the causes was deterred until the development of the X-ray machine permitted the investigator to see the phenomenon he was studying. In the case of racial and nationality differences in specific mental or manual skills, it was long (and still popularly) believed that significant genetic factors existed to differentiate racial or nationality groups, and that such factors were responsible for the apparent differences. Today we know that it is only the incidentally related factor of differing cultural emphases upon particular skills that accounts for such group differences.

IX.15 Often, of course, the relevant variables are not found simply because they are unsuspected. (This defect, it should be recalled, is a consequence of a faulty or unimaginative hypothesis.) A famous example of this possibility would be the case of bacteria as a cause of fermentation—a cause unsuspected before Pasteur's dramatic discovery and demonstration. Previous investigators had not suspected the possible existence of microscopic organisms as causative agents; and therefore did not look for them. As a matter of presumably historical fact, several distinguished scientists even refused to look into a microscope trained upon a specimen of bacteria! A comparably famous example would be the case of Fleming's discovery of the medical possibilities of a certain mold ("penicillin"). Generations of investigators had noted the toxic or inhibiting effect of certain molds; but it remained for Fleming's insight to suspect that they could be synthesized as antibiotics and thus play their present often crucial role in medicine.

IX.16 As far as controlling variables is concerned, it should be understood that the term "control" does not necessarily imply a static condition. A controlled factor may be one whose attributes do not change or are not permitted to change; but it also may be one whose attributes change but in ascertainable amounts and/or under known conditions and/or in perceivable directions. For example, if the increasing age of a boy (the control factor) is being related to his increase in height (the variable), his age certainly is not static; but its amount of change can be known and therefore compensated for in a variety of ways according to the demands of the hypothesis. In the case of studying the effects of two different types of wood preservatives upon a given species of lumber under given conditions of weathering, the reaction of the preservative upon the particular type of lumber might be controlled and the conditions permitted to vary, or the conditions might be stabilized (controlled) and the reaction to the lumber be permitted to vary. In short, control simply denotes a determinable quality or effect.

IX.17 Any number of attributes (weight, size, color, etc.) may be present in the two phenomena being studied for relationship; but it is essential that the various *relevant* attributes under control be equated in both cases. Thus, for example, if a lump of coal is being tested to determine its length of burning, it is necessary—in testing the constant (in this case, according to the hypothesis, its mass) against the unknown variable (viz., the length of burning)—to control all other attributes present (e.g., color, strata, shape, impurities, oxygen content of the air) which might possibly influence the variable. Here particularly is a most baffling problem for the social sciences; for, unlike inanimate phenomena, humans often react differently when they suspect that they are being used as controls. A competent research design in social science, therefore, must always try to employ such techniques as might overcome this always present hazard. Such techniques vary with the type and size of the sample, the instrument employed, and the behavior being investigated.

IX.18 In practice variables are controlled either by manipulation (i.e., holding them constant) or by measuring their specific influence. The former method is obvious; the latter requires various statistical techniques too complex to consider here. More often, however, the two groups (i.e., the controlled and the test) are equated by the techniques of (a) precision matching, (b) matching by frequency distribution, or (c) randomization. Each of these methods has its particular appropriateness, depending upon the kind of phenomena being studied and the degree of accuracy desired; but the three methods are not mutually exclusive, and may actually be combined for greater accuracy. *Precision* or *pair matching* means simply that the controlled attributes of one sample are equivalent to the corresponding attributes in the other sample. Thus in the illustration above, the shape, color, strata, impurities, etc., of two different pieces of coal would be matched so that the only significant variable left in the experiment would be the difference in weight or mass.

IX.19 Controlling variables by *matching by frequency distribution* means that where several different quantities of a given attribute are present, the average of those quantities is employed in the two samples being studied. Matching by frequency distribution is based upon the premise that if the averages and the distributions of two groups are similar, then one group is deemed to be representative of the other. Though in practice this method is simpler than precision matching, it has the major disadvantage that differences in the extremes between the two groups are cancelled out by the process of taking their average; and in some instances such differences might be very significant. For example, two groups may both average five-thousand dollars a year in income; but one group may have only very high and very low incomes while the other may have no very high or very low incomes. Though somewhat less precise than pair matching, frequency matching is usually far easier to achieve; and may be just as useful provided that an equal representation of the two groups of attributes can be achieved, and that any extreme cases can be shown to be insignificant to the purposes at hand.

IX.20 *Randomization* is achieved by using some system (e.g., by employing tables of random numbers) of assigning subjects to the two groups to be employed in the analytic design in order to prevent bias from influencing the selection. The major purpose of randomization is to assure the elimination of selection bias, and therefore to assure that—whatever unknown variables may be present—all variables would have an equal chance to be found in both groups. Incidentally, randomization may be used either alone or in conjunction with frequency matching; but regardless of the type of matching employed, randomization should always be employed in order to control for unknown variables.

IX.21 To avoid possible confusion, it should be pointed out that there is a basic difference between random sampling and randomization. Random sampling refers to the process of selecting units for sampling purposes; and is concerned with assuring that each unit in the universe or population being studied has an equal chance to be selected for the sample. Randomization, on the other hand, refers to the process of assigning—at random

—one portion of the derived sample to the control group, another to the experimental group. It is not necessary, incidentally, that the sample be equally divided in quantity between the two groups. For any number of purely practical reasons the sample may be split according to a variety of proportions between the two groups—provided, of course, that the two groups remain representative of each other.

C. Quantification

IX.22 The second basic problem of analytic method (once causation is established) is that of determining the quantitative relationship between two or more variables. Suppose, for example, that it has been reliably established that the addition of vitamin C (ascorbic acid) to drinking water inhibits the appearance of scurvy. The question now becomes: How *much* vitamin C is minimally necessary to achieve such an effect? Or as expressed in statistical terminology, the question becomes: What is the correlation between the (average daily) amount of ascorbic acid taken in the body and the appearance of scurvy symptoms? The problem at this stage must be cast in a mathematical form to permit a quantitative expression of the degree of relationship between these two variables, (viz., vitamin C and scurvy).

IX.23 The modern development of probability theory and its related techniques of analysis has minimized this problem of quantitative determination; and skill in statistical methods is a *sine qua non* of any competent scientist involved in analytic studies. But a word of warning should be expressed at this point if only to avoid a very common misunderstanding, to wit: Statistical correlations can only express the degree of relationship between two or more variables; they cannot designate which variable is "cause" or which is "effect."

IX.24 Statistical techniques are employed with the method of concomitant variation. In this method of inductive inference a correlation is expressed as (a) "positive" (see Figure 5) if the two variables change in the same direction—e.g., increasing pressure on the gas pedal causes a concomitant increase in the speed of the car; (b) "negative" if the two variables change in opposite directions—e.g., increasing the pressure on the brake causes a decrease in the speed of the car; or (c) "curvilinear" if the two variables are related positively up to a point but negatively beyond that point (or *vice versa*)—e.g., increasing average family income is followed by a concomitant decrease in the number of children produced; but beyond a certain point (viz., upper-class status, or in the case of very rich families), family size increases with income.

IX.25 The employment of quantification ("statistical") techniques is a positive step in the direction of achieving exactness in any science. But without intending to minimize the role of statistical procedures in modern science, a few words of caution about their use and abuse might be worthwhile at this point. For instance, statistical quantities refer only to groups of items, not to the separate items individually. In other words, a conclusion reached about the probable behavior of a group is not applicable to the behavior of any particular member of that group. To say, for example, that one out of twenty persons who will cross a given intersection at a given time will be involved in an accident does not tell which particular persons will be so involved; nor that it will be every twentieth person *ad seriatum*. Such a statement of probability refers only to the combined behavior of an aggregate or group.

IX.26 Another example of this same generalization can be cited from World War II experience. In a given U.S. Air Force squadron operating in the South Pacific during a particularly heated period of battle, the casualty rate for pilots was one per day. In a given squadron, say, of twelve pilots, it might be assumed that any new pilot would, therefore, expect to have only eleven days of combat before being shot down. Such a contention would be true only in terms of all the pilots involved over a period of time; but any given pilot might (and, in fact, some few actually did) survive several weeks of such fighting without ever having been shot down. A similar mis-assumption of "chances" (i.e., probability) is made every day by those who wager on horse races, slot machines, or various games of chance that operate within a relatively fixed probability of occurrence.

IX.27   Probably the most important statistical principle of all, however, is the one stating that correlation coefficients are in themselves not necessarily indicative of a causal relation between two series of items. (A case in point might be the example of income and family size referred to earlier. Neither factor may *cause* the other; the actual cause might be something normally related to income such as length of marriage or increasing age.) It might be demonstrable that high correlations exist, for example, between an increase in cancer in the United States and an increase in gambling in Australia, or between a decrease in the use of hair nets and a corresponding decrease in the use of chewing tobacco; but such correlations would rather obviously be spurious or fortuitous (i.e., occurring by chance). High correlations may be suggestive of meaningful interrelations; but however suggestive, in themselves are not—and cannot be—indicative of causal relations.

IX.28   In summary, it might be worth reviewing the essential role and features of analytic method. If, in broad terms, the major function of science is the acquisition of reliable knowledge, such knowledge is reliable to the extent that one can depend upon it. Such dependability is most accurately and meaningfully expressed in a predictive statement, preferably in quantitative terms. And such a statement becomes even more cogent or impressive when it contains within it an expression of causal relationship. In this form, science achieves its highest level of usefulness both as a body of accurate knowledge and as an instrument for the intelligent guidance of predictable events (whether in medicine, politics or space exploration). The analytic function of science, therefore, is the determination of reliable causal relationships accurately expressed in predictive form. To this end, the analysis of data as discussed in this chapter assumes a paramount role in the entire scientific enterprise.

D. Frame of Reference

IX.29   The *significance* of causal connections—as contrasted with the determination of them—raises a basic theoretical problem for any scientist. Somewhere within the myriad possible causal relations existing within classes of phenomena, the researcher seeks to determine which ones are more significant than others. Such a determination is expressed through the theory and its resultant hypotheses which give direction to the research. (This point has been discussed earlier.) But somewhere along the line of analysis, the researcher needs to determine *how* or *from what viewpoint* he should view his data. Such a viewpoint constitutes his frame of reference. In most instances the frame of reference is clearly indicated by the theory or the hypotheses underlying the study; in other cases it is either implied or presumed to be known. In all cases, however, the analysis of causal relations hinges upon the frame of reference underlying the interpretation of data.

IX.30   The typical research scientist proceeds with several unspecified assumptions: (a) that his colleagues or critics can understand his language and interpret it as he does; (b) that they are familiar with or can find standard reference sources; (c) that they understand his mathematical processes and other symbolic communicative devices such as maps, charts, sociograms, etc.; (d) that they are familiar with the elements of common scientific knowledge including the proper usage and interpretations of tools and instruments; and (e) that they reason as he does and pursue the same system of logic that he employs. Therefore many necessary elements of a study are either implied or assumed rather than specifically stated, so that each study need not be encumbered with the related data and procedures of organized knowledge.

IX.31   In many instances, however—and especially is this true in the social sciences (although not entirely absent in the physical sciences)—a study may proceed with one or more assumptions which are thought to be, but actually are not, established beyond a reasonable doubt. These are the so-called common sense assumptions which have littered the road of objective science. By and large, these often fallacious assumptions are of three kinds: (a) those held by most members of a specific culture at any given time; (b) those held by scientists in general; or (c) those held by the researcher alone or in common with other members of his "school" of thought. In many instances in the history of science

these common sense assumptions have been proved to be wrong—
e.g., that the earth is flat, that fire cannot burn without "phlogiston," that heavy objects necessarily fall faster than light ones, or
that personality is inherently related to race or geography. It is
imperative, therefore, that every research study make clear either
implicitly or explicitly the essential assumptions involved in its
design. Such assumptions, for example, may be made in respect
to the validity of the instruments employed, to the size of the
sample utilized, to the definitions of the terms central to the
research design, or to the statistical formulae employed.

IX.32 In some instances, however, more than simply basic
assumptions is involved; or, to put it another way, the assumptions may actually be either biases or prejudices. That is, the
researcher may knowingly or unknowingly favor one side of what
he does not recognize as (or refuses to admit is) a controversial
question. What he assumes to be obvious or proved may not be
regarded as obvious or as proved by others. Pseudo-science fails
particularly in this respect, as evidenced by the case of (a) water
witching—which contends that underground water can be located by the use of a divining rod; (b) phrenology—which contends that a person's character can be determined by the shape
and protuberances of his skull; and (c) racism—which contends
that personality is a natural consequence of one's racial attributes.

IX.33 Legitimate science has also suffered at times from this
defect of operating from a set of incorrect assumptions thought
to be true. Examples—each of which shall be taken up and
elaborated in turn—might include (1) the organic bias in psychiatry, (2) the instinctivistic bias in psychology, and (3) the
ethnocentric bias in anthropology. In the case of (1), the organic
bias in psychiatry long favored the view that mental disorders
were a consequence of purely anatomical or physiological causes.
In the case of (2), the instinctivistic bias in psychology long contended that behavior was solely a consequence of inborn, unalterable and often compulsive tendencies. In the case of (3), the
ethnocentric bias in anthropology led most Western European
and American anthropologists to view simple cultures as "primitive" or "aboriginal" or even "savage" (rather than as simply
different), thereby reflecting an unscientific attitude of cultural

snobbishness and provincialism ("ethnocentrism"). Therefore it is the duty of a scientist to specify clearly what his biases are (i.e., what his frame of reference is). This means, in effect, that he should be cognizant of what concepts, measures, units, etc., are commonly accepted by the respected members of his field; and if he should employ any that are not so accepted, should explain his reasons for so doing.

IX.34 Of course if a scientist simply does not recognize his biases for what they are, then he is vulnerable to criticism by his colleagues. It is from such instances that disagreement often arises among scientists and in turn leads to the development of differing "schools" of thought—just as differences of viewpoint lead to factionalism among laymen, who quite often defend strong biases or prejudices as though they were proven facts. But in science such disagreements eventually tend to be resolved by an empirical demonstration of objective facts. It is for this reason that the persistence of differing schools of thought among scientists demonstrates the existence of either (a) a new field of inquiry handicapped by inadequate theory or facts, or (b) a field which is changing rapidly due to the introduction of new theories or facts. When a school of thought arises outside an established field of science, then it may persist (and might even grow) because it has rigidified itself into a cult. Examples of such cultish fields operating on the fringes of reputable sciences are food faddists and old age rejuvenators.

IX.35 There are occasions, however, when a scientist admittedly advocates a legitimate bias in his choice of hypotheses or in the analysis of his data. This bias, or frame of reference, orients him to his data; and thus satisfies the requirement of objectivity in those instances where honest and competent scientists differ in their interpretation of reality. In the majority of cases, however, a specific frame of reference is employed not for the reason just stated, but simply because the researcher has a significant and logically defensible reason for emphasizing one phase of his problem more than another.

IX.36 The difference between the two reasons for a biased approach mentioned above often creates unnecessary confusion in the minds of critics. By way of illustration, an example of each might be helpful. In the first case, imagine two researchers interested in the problem of physical health. Assuming equal knowledge by both, let it be further assumed that it is not definitely clear at present whether, in the choice between diet and exercise, one is more influential upon health than the other. Each researcher, therefore, might very legitimately proceed to study the influence of one of these two factors to the exclusion— but *not* to the denial—of the other; and both would be pursuing a very meaningful scientific objective in a very legitimate manner. In a second example, imagine a researcher interested in the problem of improving worker efficiency. According to his knowledge and viewpoint, the most significant factor which contributes to worker efficiency is, let us say, morale (*versus,* let us continue to assume, such other possible factors as diet, color of machines employed, speed of the assembly line, or previous employment). He will therefore proceed to study all the factors deemed relevant to the improvement of morale, and ignore—but *not* deny— that another researcher might legitimately and profitably study the same group or phenomena for a different reason or purpose (e.g., lowering the cost of production, improving race relations, rehabilitating ex-convicts, or integrating the plant with the local high school shop program).

IX.37 In order to guard himself against possible distortion due to bias or ignorance, the scientist strives for objectivity in two major ways. First, he checks and re-checks his assumptions, his instruments, his data, his references and his inferences until he feels reasonably sure that he has overlooked no possible source of error. Second, he seeks corroboration of his methods, observations and deductions by respected colleagues, since he knows that scientists are always interested in helping each other in the pursuit of reliable knowledge. If he should differ with his peers, then he must decide which group seems to be more reasonable—or whose arguments seem to be more cogent—than the other; and then proceed until he obtains a verdict of empirical verification—a critical topic discussed in the following and final chapter.

CHAPTER X

Interpretation and Verification

A. INTERPRETATION: BASIC PRINCIPLES

B. Verification: The General Problem

C. Verification: Basic Principals

A. Interpretation: Basic Principles

X.1   The interpretation of data is determined by the demands of the hypothesis. In a descriptive study accuracy and pertinency are the only demands made of the data, and the interpretation is inherent in the factual question. Thus to ask: "What will be the difference in rainfall between New York and San Francisco during the next year?" requires simply an objective measurement of the appropriate phenomena, with practically no interpretation involved—other, perhaps, than that of defining the terms and of specifying the methods to be employed. In a simple analytic study, a relatively uncomplicated hypothesis may virtually dictate the interpretation of the data. For example, the hypothesis: "More pedestrian accidents occur during rainy weather," requires—after defining the terms and stating the limitations of the conditions—a relatively simple level of interpretation.

X.2 In more complex analytic studies, however, interpretation involves a more elaborate form of inference. As was discussed in the previous chapter, the varieties of relationships possible between two or more phenomena permit various interpretations—more than one of which could be valid, depending upon the questions asked and therefore the answers sought. The essential problem of interpretation, then, is that of employing correct inference, not only in relation to the design, but also in relation to every phase of the total study. Apart from the ever-present possibility of making mechanical errors in computation —or perhaps even of losing large batches of data—the outstanding errors of interpretation are those consequent upon the abuse of basic principles of logical inference. Since logical errors of data interpretation appear occasionally even in reputable studies, it might be worthwhile at this point to consider in summary the more basic principles of data interpretation and the concomitant errors arising from the abuse of such principles.

X.3 The principles to be discussed below—and their concomitant abuses designated as errors—summarize the most significant features of inductive and deductive inference treated earlier in various chapters. These particular principles represent the application of logical inference to the specific problem of data interpretation inherent in an analytic study design. In this sense they may be viewed as integrative instruments which tie the hypothesis, the data, the study design and the methodology together into a meaningful whole. Since they combine within themselves several subprinciples of logical reasoning they are offered here only as check points of analytic interpretation. Stated briefly, they may be summarized as follows.

X.4 *(1) Look for meaningful implications of the hypothesis.*

Since most hypotheses may have several implications for research, it is important to find those which will (a) lead to fruitful avenues of research methodology, and (b) result in a significant contribution to knowledge. Thus, for example, the hypothesis that combustion may occur even in the presence of moisture might suggest the relatively insignificant implication that the underground roots of a tree can be made to burn; but a more meaningful implication might be that, given a supply of available oxygen, a gas might be ignited even under water, thus permitting the development of a torch to cut steel ships.

X.5 Selecting meaningful implications requires a good deal of insight, but also a thorough grasp of the relation between a hypothesis and its related theory. Very often a hypothesis is viewed only in its narrower implications, and therefore its full potentiality is unrecognized. The history of invention and discovery clearly illustrates this fact. In the case of gunpowder, for example, the hypothesis that the rapid burning of certain combinations of chemicals could induce an explosive force was viewed, in ancient China, only in its relation to creating noise-makers. The more meaningful implication of this hypothesis—at least historically speaking—has been that such a force could be employed to hurl projectiles, and thus stimulate enormously the development of modern firearms. The consequences to world history are obvious.

X.6 *(2) See that the hypothesis being tested is logically related to its theory.*

Too often the hypothesis selected for testing is not logically related to the implications inherent in the theory from which it is derived, even though it may appear to be so under superficial examination. For example, the theory that women are organically related to men might suggest the hypothesis that therefore their bodily functions are essentially similar. Yet this implication ignores the fact that in some instances the sexes also may exhibit unlike functions and attributes of even greater (perhaps, in this case, of social) significance. The many myths in folklore related to this particular error in reasoning about either actual or presumed sex differences are too numerous to bear repetition. In such cases, the fault lies in the fact that the hypothesis does not adequately represent the conditions and limitations inherent in the implications of the theory.

X.7 The logical relation between a theory and a hypothesis is not always clearly evident. In fact, in some instances the relation may appear quite tenuous, whether it is logically so or not. Of great historical significance can be cited the example of the evolutionary theory of biology proposed by both Wallace and Darwin. Among the immediate reactions to this theory were the hypothetical arguments that this theory denied the validity of *Genesis*. Yet the theory as expounded by Wallace and Darwin did not necessarily infer any particular hypotheses about man's origin as far as the "original pair" notion of *Genesis* is concerned. Years later—after some of the original fury of debate had subsided—it became obvious to critics on both sides of this argument that the evolutionary theory as such could logically suggest several different hypotheses of man's origin; depending upon how literally one interpreted the Wallace-Darwin statement, and how literally one interpreted the account in *Genesis*. In short, the logical relation between a theory and one of its derived hypotheses is not always clearly evident; and when it is not so, then it is incumbent upon the scientist to make the relation between the two clear. In many instances the simplest way to do so is to cast the argument in the form of a series of syllogisms of continued reduction. Thus, by adhering to the principles of a valid syllogism, the relation between a theory and a particular hypothesis can be established through deductive inference.

X.8 A theory, in other words, is usually stated in general terms and therefore without qualifications; for it expresses only an abstract situation operating under theoretical conditions. To state the theory (sometimes called the "principle," or the "law") of the pendulum (viz., that the timing of its swings is related to its length by the law of inverse squares) for example, is to ignore— but *not deny*—the fact that this theoretical relation never obtains in reality. The reason, in this case, lies largely in two factors: resistance imposed by mechanical friction, and that imposed by air currents. The research design, therefore, should test those hypotheses which are logically related to and are verifiable consequences of the theory. In the example of sex differences referred to previously, the design should be so devised to prevent overgeneralization of the conclusions; otherwise the theory might be interpreted to imply a large degree of generality—and thus suggest hypotheses of high generality. Whenever in doubt about the relation between a theory and its various possible hypotheses, caution suggests the employment of continually refined (i.e., increasingly discriminating) four-celled designs. In this way the relationship between theory, hypothesis and design will of necessity become increasingly clarified.

X.9 *(3) Be aware of the possible influence of unknown variables.*

The ever present danger in this case is not one of faulty logic (as are the cases referred to above), but of lack of recognition of the type of causative relationship. The difficulty here is generally due to faulty knowledge, and therefore to faulty design. If, for example, it has been decided to test the influence of nutritional factors upon two different racial groups, the possible differences derived may erroneously be attributed to the controlled variable of race. But as any student of nutrition knows, nutritional reactions are not simple discrete factors. Rather, they are a complex consequence of individual, social and chemical factors which influence the total bodily responses of the organism. In this case, the differences in nutritional response—though different for the two racial groups under study—might be a consequence, not of the organic factor of race, but of the social (i.e., dietary) factors incidentally or coincidentally related to the two groups being tested.

X.10 Various checks can minimize this problem of extraneous variables: the employment of a "null" hypothesis (i.e., a statistical test to indicate whether the phenomenon could have occurred by chance alone), the use of random sampling and of randomization, the use of control groups, etc. In the last analysis, however, it is almost impossible to ascertain a clear and simple cause-to-effect ("invariant") relationship in any complex kind of research phenomena. What this principle suggests, therefore, is more a note of caution against deterministic interpretations. If, as often happens of necessity, one is approaching a problem symptomatically, then he is not too concerned with this principle; for it doesn't matter what the "real" causes of a problem are as long as its effects can be modified. Thus the physician may not be too concerned about the influence of unknown variables *causing* a headache if his major problem is to *alleviate* it. The aspirin, let us surmise, in this case apparently operates upon the significant variables (whatever they may be) causing the headache; and the palliative—in this case the aspirin which effects the symptoms of the headache—certainly serves a useful purpose in medicine. But if one is interested in finding the basic (i.e., necessary and/or sufficient) causes of the headache, then a symptomatic approach is practically useless; except perhaps to suggest which variables may or may not be suspected as related links in the chain of causation.

X.11 (4) *Be sure that the sample and its universe are legitimately related.*

Wrong attribution between a sample and its universe is a very common error of many research studies. This type of error occurs when the results of a study made upon a specific sample are extended to the whole class from which the sample was drawn, thereby ignoring the probably meaningful differences present in the total universe. Suppose, for example, that a study indicated that a sample of sandy soil grew better plants than did a sample of rocky soil. This conclusion would be logically applicable only to that type of soil under those conditions and for those particular plants; different types of sandy soil under other conditions or for other types of plants might have just the opposite consequences.

X.12 The converse form of the error of wrong attribution referred to above also occurs frequently. Suppose, for example, that it was found that auto accidents occur more frequently in winter than in summer. It would be erroneous to infer therefrom that on any given day or night in winter more auto accidents would occur than during corresponding days or nights in summer. The generalization in this case would be valid only for the seasons as a whole, but not necessarily for each given day of each season.

X.13   (5) *Check the qualifications for limited implications of confirmation.*

The lack of clearly designated qualifications necessarily leads to either overgeneralized or undergeneralized implications of the confirmed hypothesis. Of the two extremes of implication, over-generalization is by far the more common form. This too-common error invariably results from inadequate sampling, and can be avoided by simple statistical techniques. Suppose, for example, that it was found that mathematics majors were disproportionately blondes—i.e., compared to physics majors, or to music majors, or to the student body at large. It certainly would be erroneous to assume therefrom that hair color is somehow related to one's choice of an academic major field—unless, of course, large enough samples were employed to rule out the probability of chance occurrence. Laymen are notoriously susceptible to this type of erroneous reasoning.

X.14   Another version of the abuse of this principle results from generalizing beyond the limits of the research design. This type of error extends the results of the study into fields similar to but not identical with those studied. Suppose, for example, that a study of corrosion resistance established the finding that aluminum outlasts galvanized iron. Though generally true, this finding would be erroneous when applied to situations near the seashore, where the action of salt water in the atmosphere is particularly corrosive to aluminum. Of all the errors committed in the name of science, this general type probably overshadows all others combined; and in spite of the possibility of undergeneralizing, should be guarded against most assiduously.

X.15   The converse form of overgeneralizing is the error of undergeneralizing. This type of error is usually found among overly cautious scientists, and consists of the tendency to demand unequivocal or absolute proof before a hypothesis can be considered confirmed. Suppose, for example, that a small relationship has been found to exist between certain political events and retail sales. Though small, the relationship might still be very significant to the retailer who is responsive to political events. Fortunately, there are statistical techniques which permit calculation of the degree of significance of a relationship; and it is common practice to express correlative relationships with the expression "at the five- (or ten-, or one-) percent level of significance." Unfortunately, however, there is no general agreement within some scientific fields as to the level of significance deemed necessary or desirable before a given relationship can be considered meaningful.

X.16 The reason for this lack of agreement inheres in the two different implications of the term "significant." A relationship is deemed statistically significant when it can be reasonably assumed to be operating beyond pure chance (or "zero") probability—i.e., when it can be presumed to be somehow related beyond pure coincidence. To be significant in a non-statistical sense, however, a relationship—as in the previous illustration of political events and retail sales—derives a meaningful implication only in terms of how it relates to the hypothesis being tested. It is this type of possible lack of agreement that often creates differences of opinion and even disputes among scientists. In short, statistical and theoretical or functional significance bear no necessary relation to each other.

X.17 *(6) Check the implication of negative cases.*

Often a dramatically vexing and stubborn question of interpretation is posed by the presence of a negative case or exception to the general rule. Errors resulting from ignoring the exceptional case have often highlighted the slow growth of modern science ever since its beginnings. The role of the negative case poses an important problem of interpretation. Contrary to the popular expression, the exception does *not prove* the rule either in logic or in science, it *invalidates* it. The negative case (i.e., the one that does not fit the concluded generalization) suggests either of two possibilities: (a) that the conclusion is overgeneralized insofar as deviations or exceptions have not been accounted for; or (b) that the conclusion is substantially incorrect. In both cases the fault may lie either in the design or in the hypothesis, but generally in the latter; for invalidating exceptions usually result from definitions that are too broad or from causal relationships expressed in too large a degree of generality.

X.18 The problem of the exceptional case should always serve as a signal of caution to the alert researcher. Logicians of scientific method have long debated the significance of the so-called "crucial experiment"—i.e., one that definitively confirms one of two contradictory hypotheses. In this same context the question often arises as to whether an exception necessarily invalidates a general rule, or even whether only one of two different (and perhaps even contradictory) hypotheses can be deemed valid. (One facet of this problem was treated in Section A of Chapter IV in relation to operational definitions. The question raised there was: Can two different operational definitions be said to be designating the same phenomenon?) For the purposes of this book, this fundamental question probably can best be answered by saying that any phenomenon can be adequately explained by various hypotheses, depending upon how the theory is employed—i.e., upon one's frame of reference. Since science is a synthetic as well as an analytic system of deriving knowledge, the role of the negative case should be viewed in terms of the logical implications of the theory underlying the hypothesis. This principle of interpretation, therefore, needs to be viewed in its wider application.

X.19 (7) *Look for meaningful differences within classes of phenomena.*

The error in this case results from the common tendency to feel that the general differences between two groups of variables are meaningful while ignoring the equally important fact that the individual differences within the groups may be even more meaningful. Thus, for example, the general differences in athletic performance or in driving ability between men and women can easily lead one to the conclusion that men are decidedly superior to women in either athletic or driving ability. This conclusion ignores, however, the very meaningful fact that many women are significantly superior to many men in both athletic and driving ability.

X.20 The scientist, metaphorically speaking, is constantly walking a fence. On one side is the principle of parsimony, which encourages him to explain as much as possible with as few formulations as possible. This is the generalizing tendency inherent in all theoretical science. But on the other side of the fence is the principle of exactness, which demands that he employ clear and unambiguous concepts and constructs. This dualistic attitude is especially appropriate to this particular principle of interpretation. On the one hand, the scientist prefers to generalize as far as possible, but on the other hand, does not want to generalize beyond the limits of his hypothesis and data. (These two desires are not contradictory.) This principle therefore suggests that he exploit as far as possible all the meaningful implications inherent, not only between classes of phenomena, but also within those classes as represented by the differences between subclasses. Thus, to refer to the example above, the psychologist who is interested in the difference either in athletic performance or in driving ability between men and women should not, according to this principle, ignore the significant differences among women and among men—as well as between some women and some men, and between women and men as whole classes.

X.21 *(8) Denote attribution (causation) clearly.*

The question of causation is—as has been noted several times previously—of paramount significance to the scientist. Even though he views causation only in terms of antecedents and consequences, nevertheless he cannot escape causal implications. Indeed, he is basically concerned with ascertaining the direction —as well as the force, frequency, and consistency—of such implications. The error of wrong attribution, then, results when factors are related to the consequences but not in a causative manner. If, for example, it was found that wars occur disproportionately during one particular type of political administration, it might be assumed therefrom that the type of administration in power is the determinant factor leading to wars. The error here is the neglect of the "real" causes of war (whatever they may be) which might have been, in this example, only coincidentally related to the type of political administration found in office at any given time. Within this class also are found those interpre-

tations made to fit prejudices, strong biases or other types of mis-
conceptions. Most errors of attribution are of the *post hoc* variety,
the form of thinking most commonly employed by laymen. A
common example is found in psychotherapy, where many "cures"
are attributed to widely different theories of causation. But since
there is at present *no* (demonstrably) *necessary* relationship between the efficacy of any particular psychotherapeutic system and
the rationale of its supporting theory, it is not surprising to find
that cures are effected by totally differing methods and rationale
systems—e.g., holy shrines, psychoanalysis, patent medicines, special diets or exercises, incantations, pills—all of which probably
operate by sheer suggestion alone.

X.22   The outstanding problem of interpretation, however, is
the assignment of causation. As has been discussed previously,
the scientist today prefers to answer the question of causative relations simply by expressing a statement of probability sequence
(e.g., when B occurs it is preceded by A to a certain degree of
probability). But in some instances the biased researcher attempts
to explain such relationships by indicating causative factors of
teleological implication rather than by being satisfied with probability statements alone. Logically, for example, there would be
a vast difference between the two statements: "Fire is caused by
striking a match"; and "Fire follows the striking of a match in
such a percentage of cases when so-and-so conditions obtain." In
the second statement the implication of teleology is avoided, and
thus represents a more cautious and therefore more logically
defensible interpretation of the demonstrable sequential phenomena.

X.23   In all interpretations, caution, intellectual modesty and
skepticism are fundamental guards against most interpretive errors. Instead of asking oneself "How much did I prove?" the
exacting scientist asks himself "What is the least I have proved?"
It is for this reason of "least proof" that the scientist prefers to
begin the analysis of his data by subjecting them to the test of
the null hypothesis, as mentioned earlier, in order to be assured
that at least simple chance or coincidence was not responsible for
the results obtained. Always dubious, the critical researcher tries

to anticipate all logical criticisms of his conclusions as well as of his design and methods. He checks especially for "particularism" —i.e., the tendency to ascribe consequences to single causes— the so-called "single-factor fallacy" mentioned earlier, and sometimes referred to as the "particularistic fallacy"; and he often assumes the role of "devil's advocate" by challenging his own conclusions. Only when he can no longer conceive of any reasonable objections to his conclusions is he satisfied that he has gone as far as his skill, knowledge and data permit.

B. Verification: The General Problem

X.24   The whole effort of a scientific study is directed toward one single end: verification. In a descriptive study verification consists of corroboration of the expressed results, generally by replication of the observations by unbiased observers. How many replicate observations need to be made, and how great a degree of difference among the various observations is to be tolerated, however, are questions that can be answered only by reference to the hypothesis. In an analytic study verification consists of the prediction of sequential relationships expressed in terms of degrees of probability under given conditions. If a study has been well designed and well executed, then the problem of verification is vastly simplified. If, however, the problem has been loosely formulated, if it has not been made clear which data are pertinent, if all relevant variables have not been accounted for, if the methodology has been inadequate or inept, if the techniques were inappropriate to the data, or if the conditions of analysis were such that empirical demonstrability is impossible—then verification cannot be achieved objectively.

X.25   In spite of many variations, there are only two basic types of verification accepted in legitimate science. The basic and optimal form is prediction beyond pure chance; the second, consensus. In the first instance confirmation of the hypothesis—and of its establishment therefore into a theory, a principle or a law— is effected by demonstrating a sequential consequence operating under stated conditions of significant and designated probability.

X.26   It should be remembered, however, that "proof" can be established by logical deduction as well as by verification of a probabilistic statement. Proof in a logical sense is a matter of establishing a sound deductive argument (as discussed in Chapter II). What is here referred to is essentially the translation of a logical argument into its empirically testable equivalent by substituting specific (empirical) terms for the qualitative terms of the argument. These two forms of "proof" (i.e., logical deductions and probabilistic statements) are not, however, synonymous; nor is the one always or necessarily "translatable" into the other.

X.27   The terms theorem, principle and law have undergone some changes of interpretation and usage in recent years. Historically, a *theory*—as distinguished from an hypothesis—referred to a formulation of apparent causal relationships existing between certain classes of observed phenomena, and which had been verified to some degree. A *principle,* on the other hand, referred to a fundamental truth, doctrine or motivating force; and as such was synonymous with the concept of so-called *natural law.* Today, however, the historical distinction between a theory and a principle, or between a theory and a law, has become less definite, due to a growing suspicion of the categorical permanency implied by the term "natural law." In social science the general term "theory" is employed almost exclusively when referring to apparently established principles of causal relationship —e.g., theories of learning, the theory of the culture concept, game theory, the culture-lag theory—while in the physical and biological sciences the term "law" is generally reserved for the well-established principles once thought to be unalterable—e.g., Newton's laws, Boyle's law, Mendelian laws, the law of evolution. (Another facet of this problem of designation was referred to in the first section of Chapter VIII.)

X.28   In effect, verifiability may be regarded as an extension of the concept of reliability—i.e., explanations of phenomena in terms of predictability in given instances; or, in other words, to evidence predictable consistency. Thus, for example, to say that the theory of inherited characteristics has been "validated" by demonstrating it in a given number of predictable instances is tantamount to saying that the expressed relationship is a reliable one. To an increasing extent, scientists tend to avoid implications of causality by thinking of verification as an expression of high reliability.

X.29 In the physical and biological sciences the confirmation of a hypothesis is usually further established by subjecting the derived principles to empirical tests. In a typical case, a small plant is constructed to test the process involved, or a small situation or sample is selected to test the principles derived. Such procedures serve the dual purpose of checking the predictive value of a hypothesis empirically, and of permitting further modification of the principles in order to increase reliability, hence validity. In such circumstances, a "pilot study," a "pilot group," or a "pilot plant" are functionally synonymous, although some small ambiguity of designation still exists in contemporary terminology. In the social sciences, for example, the term pilot study is often used synonymously with the term exploratory study. (As a reminder: an exploratory study is designed to study the efficacy of methodological procedures and techniques, and perhaps to stimulate discoveries. It is not supposed to be a verification of the hypothesis.) Yet just as often the term pilot study is used according to the implications mentioned above which are common to the physical and biological sciences. When a true pilot study is designed in the social sciences, however, it serves the same purpose that it does in the physical and biological sciences: viz., as an empirical test of the verified hypothesis as expressed in terms of a specific degree of reliability under stated conditions. At this point, reliability and validity merge, not as absolute concepts but as qualified predictions of probable effects.

X.30 The confirmation of an hypothesis by subjecting the conclusions to empirical tests in a pilot study is a familiar process in the physical and biological sciences. One has only to recall the famous instances of the pilot reactor under the bleachers of the Chicago stadium where controlled atomic fission was first demonstrated in accordance with the predictions made by now-famous physicists; and of the later verification by the first test bomb exploded as predicted at Alamagordo, New Mexico. An almost equally spectacular example was the famous pilot study made to verify the efficacy of the Salk vaccine for paralytic poliomyelitis —a study which demonstrated that in the pilot group the reliability of the vaccine's predicted effect was actually higher than initial studies had cautiously suggested. In the social sciences a less familiar example would be a series of studies made of United States soldiers in World War II, which predicted several consequences later verified empirically in pilot studies: viz., the effects of integrating Negro and White troops in certain mixed units, factors contributing to combat efficiency, attitudes toward a system of discharging men from the armed services, methods of assessing candidates for strategic services, etc. In all probability, the employment of pilot studies in the social sciences will eventually become almost as common as in the physical and biological sciences; and at that time will better serve to differentiate good research designs from poor ones.

X.31 The second type of verification—admittedly of a lower logical and functional level than the first one—relies upon consensus. Essentially, *consensus* is supported by derived authority, and authority is a reflection of established prestige. In this respect verification by consensus as employed in science is not basically different from the type of verification employed in religion or politics—except for the important difference that authority in science is admittedly objective, temporal and fallible. Consensus implies agreement in definition or interpretation, and may involve either a small or a large group of authority figures. Since authorities can easily disagree in the absence of empirically demonstrable and predictable facts, it is not surprising that verification by consensus is often highly questionable. In fact, some of the outstanding differences between the physical and the biological sciences, on the one hand, and the social sciences on the other, are often attributable to this essential difference in the criteria of verification. The same degree of difference is also operative between the social sciences on the one hand and the humanities on the other.

X.32 Although substantially inferior to verification by prediction, verification by consensus represents an intermediate level of scientific achievement, and as such should not be scorned in the absence of anything better. Yet sound reasoning dictates suspicion of the common notion that "something is better than nothing." There are too many instances from the history of science as well as from politics, law and economics demonstrating that this notion can lead to disastrous or even fatal consequences. The common experience of mobs—who because of fright or anger choose to "Do something!" (often with disastrous results)—well illustrates the dubiosity of the assumption that something is necessarily better than nothing. In politics, law and economics the history books are filled with examples of man's disasters resulting from erroneous notions applied in reaction to attitudes of helplessness or futility when a people were faced with unsolvable problems. In the field of science, one has simply to note the innumerable instances where the acquisition of verified knowledge was diverted and thereby long delayed because impatience caused the acceptance of unverified hypotheses. The previous statement—that verification by consensus should not be scorned in the absence of anything better—therefore should be construed reservedly.

X.33   As employed in science, consensus is generally one of three types: (a) agreement among the established and accepted leaders in a given field (authority figures), (b) agreement among groups of authorities as defined by their followers within a select group ("schools of thought"), or (c) agreement by virtue of isolated declaration by a single person. Generally speaking, a higher degree of validity tends to follow the order mentioned here; but too many exceptions have occurred in the history of science to permit a dependable generalization to be made about this hierarchy of authority. There have been too many instances, for example, where the atypical scientist was later vindicated by the verification of empirical prediction. Examples which come immediately to mind are the case of Columbus, who was scoffed at because he insisted that the world was round; of Pasteur, who was villified for his germ theory of disease; of Harvey, whose professional practice suffered for many years because his theory of the circulation of the blood was rejected; of Einstein, against whom an organized campaign of ridicule was launched until his theories of relativity were finally accepted; and of Darwin, whose theory of evolution is still prohibited from being taught in some sections of the United States.

X.34   As far as the second level of consensus is concerned, the novice should not ignore the fact that the history of science records many instances where differing schools of thought have argued almost as acrimoniously and as pettily as two rowdies in a barroom brawl. Therefore, in cases where disputation exists between differing schools, verification must remain essentially qualified by the implications of the hypothesis—i.e., the unbiased scientist must choose among the differing groups as best he can according to the reasonableness of the hypothesis and the cogency of its confirmation as determined by his accumulated knowledge and experience. Until the differing groups or individuals can agree upon objective criteria of validation, consensus remains impossible to achieve. If, however, they do manage to agree upon the criteria of validation, all they will have achieved is consensus, not necessarily validity. Optimal validity still remains a matter of empirical demonstrability expressed in a logically defensible prediction of probable effects—a prediction which must be better (i.e., more accurate) than one achieved by pure chance alone.

X.35   An important problem which has often vitiated the social sciences—especially psychology and psychoanalysis, but even economics, anthropology and sociology—is related also to the second level of consensus mentioned above (i.e., agreement only within a given school of thought). Many instances could be cited to illustrate the fact that opposing schools not only can exist side by side but in many cases even manage to proceed with a serene assurance that their differing theories have been "proved" by demonstrable evidence and even by prediction. Two explanations account for this situation: (a) Either or both schools may be wrong in fact as established later or by other authorities, but so define their criteria of validation that the predicted consequences are interpreted to fit their respective theories. Essentially, this process is simply biased sampling, biased selection of data, or more often biased interpretation. (b) Either or both schools may later be proved to be correct in fact, but to be functioning in different areas of phenomena or, more often, to be operating from essentially different hypotheses. In most cases, however, such differing proofs are a consequence of closed systems held together by the specious logic of circular or tautological proof. To a critic, a system of circular proof is logically unassailable; and it is just as futile to try to disprove such a system which masquerades as science as it is to attack it when it appears in a form of religion, politics or aesthetics.

X.36 It is for such reasons that pseudo-science continues to exist by "proving" its specious "laws." In many cases, of course, such so-called proofs are simply outright quackery of the kind familiar to police bunco squads, or that commercial advertisers foist upon a gullible public. But in most cases, pseudo-science beyond the level of sheer fakery continues to convince skeptics (a) by operating on the level of sheer suggestibility, as evidenced by the pitchman or huckster "proving" the efficacy of his "snake oil," by the apparently magical proof of an ouija board, or by the startling "proofs" of the shaman who sticks pins into an image of a victim who sometimes does later die (but of fright, of course); (b) by combining circular proof with selected instances, as witness the "proof" of water diviners, hands-on healers, or "Monday-morning quarterbacks," or (c) by dogmatically interpreting either a cause or an effect in the face of counter reputable interpretations, as claimed by the "proofs" of Marxian economics, of "occult science," or of faith healing. In essence, then, unscientific proof relies either upon (1) the acceptance of dogmatic authority coupled with a closed mind, or (2) invalid reasoning. In either case, scientific counterproof can make no headway until both of these barriers are removed.

X.37 The problem of verification in empirical science resolves itself therefore, into four aspects: (a) the logical structure of the hypothesis and of the research design, (b) the precision and appropriateness of the methods, (c) the criteria of reliability and/or validity, and (d) the level of credibility of the investigator. Strength in one or more of these aspects does not compensate for weakness in the others; for like the proverbial chain, verification can be no stronger than the weakest link in the total research effort. The serious student of scientific method realizes that errors in design, inference or techniques tend to multiply the improbability of reaching valid conclusions; so he verifies as far as possible each step of the total effort as he proceeds, thus leaving only the hypothesis to be tested. This means, in effect, that the criteria of validation should be clearly expressed in objective and empirical terms before the study design is ever activated either in the laboratory or in the field.

C. *Verification: Basic Principles*

X.38   For purposes of synthesis and clarification it might be fruitful at this point to bring together in the form of eight basic principles the more outstanding features of scientific method previously treated in various portions of this text. Such a compilation can serve as a series of check points of verification, especially when applied to the analysis of a study as a whole. Furthermore, such a listing might be employed as a critical screen in order to check the hypothesis against the conclusions; or it might be viewed as a specification of the essential problems inherent in an evaluation of the total research effort. Whatever its peripheral values, such a listing is most useful in summarizing the critical aspects of verification; and as a summary should be viewed simply as a functional logical device.

X.39   *(1) Check the facts.*

In accordance with this principle, the careful scientist asks: (a) Are the facts almost certainly true, or are they to be seriously questioned? (b) If true, are they implied or inferred to be always so, or are they properly qualified? (c) If of dubious quality, what evidence is offered to support them, and is the evidence acceptable? (d) Are the units of study clearly and acceptably defined, and are they meaningful as facts? (e) Are the various facts, as units of study, comparable; or are they essentially different when employed in different contexts? All these and other such questions refer essentially to the problem of clear and acceptable definition. Allied to this "verification imperative" are the related questions: (f) Have the concepts been defined either empirically or operationally whenever possible? (g) Are they defined in both standardized and comparable forms? (h) Are the units of study appropriate to the purpose? (i) Have they been objectified and metricized whenever possible? And (j) are they reproducible? In short, the verification imperative to check the facts means essentially to question every facet of every concept and construct employed in the total research study.

X.40   *(2) Check the assumptions.*

More insidious in their implications than are unverified facts, are assumptions underlying a study which often escape notice simply because they are usually inferred rather than clearly implied or specifically expressed. As was noted in Chapter II, the postulates underlying the hypothetical, the methodological and the theoretical structure of any scientific effort are of fundamental significance to any analysis of procedure. Dubious assumptions which in the past have hindered the development of modern physical and biological science are now well recognized, e.g., those underlying pre-Copernican astronomy, pre-Euclidian geometry, pre-Baconian epistomology, pre-Darwinian biology, pre-Wundtian psychology, pre-Einsteinian physics. But not so clearly recognized are the many underlying assumptions inherent in much modern biological and social science. Whatever the field, the astute scientist takes for granted no more than is absolutely necessary, and checks very critically the whole assumptive underpinning upon which his research design is supported.

X.41   *(3) Check the secondary materials.*

The distinction between primary and secondary sources was noted earlier in the discussions of library and historical materials. This distinction should, however, be extended to all presumably valid materials of a secondary sort: facts, concepts, samples, computations, theories, histories, models, structures, and other sundry constructs presumed to be beyond question. This particular verification imperative needs to be followed with caution, of course, for a judicious estimate of the reliability of secondary materials must be made at some points in order to conserve effort. But the essential feature of this principle is its emphasis upon "healthy skepticism." In this sense it should be viewed as a warning signal to the cautious scientist that (to twist a metaphor), all that glitters is not true. Even a cursory examination of much contemporary scientific literature (again, especially in the social sciences) would indicate that many studies build upon previous studies of dubious validity. The exploitation of data from some contemporary studies on sex behavior is an excellent example of this point; as is the uncritical acceptance of many essentially unverified assumptions, concepts and principles derived from psychoanalytic doctrine.

X.42   *(4) Check the instruments and techniques.*

In the writer's home workshop there are four carpenter's squares all of which disagree with one another, a steel tape which is uncorrected for inside as against outside measurements, a so-called combination square which gives a fixed angle of forty-two instead of forty-five degrees, and two identical aluminum yardsticks which differ in their internal divisions of inches. Precision-made laboratory or field instruments, of course, would not be so flagrantly inaccurate, but nonphysical instruments are quite another story. The common tendency of assuming that so-called attitude scales actually measure a specific predeliction for overt behavior, that marriage-prediction scales actually can predict marriage adjustment, that "neurotic inventories" are actually that, or that unstructured or open-ended questionnaires actually can and do give objectifiable responses, is a most naive assumption. This principle, therefore, refers primarily to the nonphysical measuring instruments employed in the social sciences.

X.43   Methods and techniques also should not be taken for granted. In the physical sciences many methods and techniques have been validated repeatedly; so there is no logical reason to question them each time they are employed. In the biological sciences likewise a high degree of reliability can be imputed to many standardized tests and techniques. Unless the scientist has previously satisfied himself that given instruments, methods and techniques are valid enough for his purposes, he should be wary of assuming that the professed reliability or validity of an instrument, method or technique has been established beyond all reasonable doubt. Of paramount importance is the replicability feature of scores or measures derived from any type of instrument. So-called interpretive tests certainly lack this feature, as do all other nonobjective, inexact or nonempirical devices employed in many fields of science.

X.44 (5) *Check the frame of reference.*

The validation of a descriptive study rests solely upon the verifiability of its facts. In an analytic study, however, the type and degree of significance of the demonstrated relationships depends upon the frame of reference (i.e., the point of view in terms of interest, or theoretical bias) of the investigator. If the frame of reference is clearly stated and effectively justified, no logical criticism can be made of the conclusions drawn from the data; provided, however, (a) that the study design, the methodology, the techniques and the data are all objectively verifiable in their own right, and (b) that the conclusions logically follow from the inferred or implied premises and from the theory underlying the study. On the other hand, if the frame of reference is not admittedly and clearly stated, then the whole study as well as the conclusions should be suspect. Inherent, often subtle, and usually unadmitted biases are found in studies of all types and in all fields; and the critic should be aware of the problem posed by this validation imperative. If the critic should himself share the biases, or operate from the same frame of reference as does the person doing the study which is being examined for validity, then obviously objective criticism will be negated or vitiated.

X.45 (6) *Check for logical errors.*

As has been noted in several chapters previously, logical errors can occur at almost any point of a study—from its beginning in terms of ambiguous definitions to its conclusions in terms of invalid deductions. Many purportedly valid studies abound in vague generalizations, omnibus terms, false dichotomies, questionable authorities, overgeneralizations or imperfect analogies. Even more difficult to ferret out, however, are the insidious errors of inductive and deductive inference resulting from poor logical designs. The only incisive tool of objective criticism is a thorough knowledge of logical methods in all their meaningful applications; for it should be borne in mind that many otherwise competent scientists often lack a thorough grasp of the intricacies of logical inference. No one type of logical error probably can be said to be more significant than another; but the history of invalid science might suggest that the most common type has been the inferences drawn from a false analogy. This error is particularly acute in the so-called behavioral sciences.

X.46   (7) *Check the analytic design.*

As was noted in the previous chapter, the optimal form of proof by inductive inference is the basic four-celled design. But a common notion exists among many purportedly competent scientists: viz., that, being "eclectic," they can legitimately choose *any* design that "gets results." Such reasoning leads invariably to the employment of methods of lesser validity than does the controlled, four-celled analytic design. This notion is notoriously prevalent among historians, but is almost as commonly prevalent among social scientists in general, and is not unknown among biological scientists. If eclecticism is interpreted to mean the willingness to employ any method that "gets the best results," then no sensible argument can be made against it from a functional point of view. But if eclecticism is interpreted to imply that any method that gets results is good simply because it gets results—or is as good as any other method that gets results, or is good enough because there is nothing better—then the critic should immediately exercise his skepticism. True comparability rarely exists in fact. It is highly dubious that one type of design is of comparable validity to another.

X.47   (8) *Check the interrelation between hypothesis, theory and design.*

This outline has stressed that, of all the manifold phases of scientific effort, the hypothesis and the study design bear the greatest load of responsibility for deriving valid conclusions. The interrelation between the hypothesis and the theory from which it is derived, and the relation of both to the study design, form a logically integrated whole whose fruitfulness in terms of valid and meaningful conclusions can be assessed only in terms of the logical consistency which the integrated whole displays. Too often in quasi-scientific studies—and always in pseudo-scientific efforts—this logical interrelationship is absent, or at least is not clearly evident. The all-too-human desire to achieve desired results often leads to questionable formulations which rest upon hypotheses not clearly related to a reasonably demonstrable theory of phenomenological relationships. The critical scientist needs, therefore, to check the overall pattern of the entire study; first in terms of its internal components, and second in terms of its pattern as an integrated whole. Only when he is satisfied that both the parts and the whole are factually and methodologically acceptable can he feel reasonably confident that the conclusions are valid. But if any aspect of the total design is dubious, and especially if the hypothesis, the theory and the design are not logically and meaningfully interrelated, then he should question the validity of any and all conclusions drawn from the study.

X.48 When one reflects upon the implications of the above listing of principles, it perhaps becomes clearer why so much poor research is done in the name of science. Faulty technique in itself (a subject not treated in this text), accounts for much incompetent research; but just as much can be accounted for by faulty interpretation inherent in the study design. It is for this reason that the competent scientist trains himself diligently in all phases of logical inference, and applies such training to all phases of a study. For if the logic of inference is faulty, precise and even complex techniques involving elaborate instruments and highly involved computations are in themselves useless for answering the questions posed at the outset. The implications of the data rest essentially upon the interpretations placed upon them; and no tools or instruments have yet been devised which can check for, let alone correct, errors in human reasoning.

Bibliography

BEVERIDGE, WILLIAM I. B., *The Art of Scientific Investigation,* London: William Heinemann, Ltd., 1950, (172 pp.).

An acknowledged British biologist discusses and historically illustrates the highlights of scientific method, particularly in the physical and biological sciences, stressing the role of chance and insight in research.

BLACK, MAX, *Critical Thinking: An Introduction to Logic and Scientific Method* (2nd ed.), New York: Prentice-Hall, Inc., 1952, (459 pp.).

A very useful, clearly written text on general logical method, with comprehension tests and exercises appended to each chapter. One of the best of its kind for laymen.

BRAITHWAITE, RICHARD B., *Scientific Explanation: A Study of the Functions of Theory, Probability and Law in Science,* New York: Harper Torchbooks, 1960, (374 pp.).

Based upon lectures given in 1946, this volume is one of the most lucid expositions ever written on the logic of mathematical statistics, of probability and its role in scientific models, and of teleological explanations in all sciences. A "must" for the serious student, but not for the novice.

BROWN, CLARENCE W., and GHISELLI, EDWIN E., *Scientific Method in Psychology,* New York: McGraw-Hill Book Co., Inc., 1955, (368 pp.).

An excellent text written for a moderate level of understanding; basic for social scientists.

CAMPBELL, NORMAN, *What Is Science?* New York: Dover Publications, Inc., 1921, (186 pp.).

An interesting and substantial treatment of various aspects of science by one of its best known writers.

CHAPIN, F. STUART, *Design of Social Experiments* (rev. ed.), New York: Harper & Bros., Inc., 1955, (297 pp.).

A good general discussion of the logic involved in experimental method, particularly as applied to social phenomena. One of the basic texts for social scientists.

248

CHURCHMAN, C. WEST, *Theory of Experimental Inference,* New York: The Macmillan Co., 1948, (292 pp.).

A very meaningful though not simple book, it treats the problems of presuppositions of inquiry, the essence of verification, and the relation between problem formulation and the value orientation of the researcher. An outstanding short text in the philosophy and epistomology of scientific research.

COHEN, MORRIS R., and NAGEL, ERNEST, *An Introduction to Logic and Scientific Method,* New York: Harcourt, Brace & Co., 1934, (467 pp.).

A basic, comprehensive and ever-popular college text treating the logic underlying scientific methods of induction, deduction, proof, syllogistic reasoning, etc.

COPI, IRVING M., *Introduction to Logic,* New York: Macmillan Co., 1961, 2nd ed., (512 pp.).

In the class of Black, and Cohen and Nagel, a very competent general college text on logic and scientific method.

FEIGL, HERBERT, and BRODBECK, MAY (eds.), *Readings in the Philosophy of Science,* New York: Appleton-Century-Crofts, 1953, (811 pp.).

An excellent anthology by many writers offering different points of view found in logic and science; including a superb bibliography.

FESTINGER, LEON, and KATZ, DANIEL, *Research Methods in the Behavioral Sciences,* New York: Dryden Press, 1953, (660 pp.).

A thorough but somewhat advanced and specialized text for social scientists, especially social psychologists.

FRANK, PHILIPP G. (ed.), *The Validation of Scientific Theories,* Boston: Beacon Press, 1956, (242 pp.).

Twenty-two papers first presented before the AAAS in 1953. Provocative articles by some of the outstanding scientists and philosophers of our time, though relatively general in treatment.

GARDNER, MARTIN, *Fads and Fallacies in the Name of Science* (2nd ed.), New York: Dover Publications, Inc., 1957, (363 pp.).

An interesting popular treatment of many past and present pseudo-scientific doctrines, their advocates and their rationalizations. A good intellectual dessert after Braithwaite, Churchman or Nagel.

GOOD, CARTER V., and SCATES, DOUGLAS E., *Methods of Research: Educational, Psychological, Sociological,* New York: Appleton-Century-Crofts, 1954, (920 pp.).

A comprehensive though not very advanced text for workers in the field of educational research, and containing unusually extensive bibliographies on materials and methods.

GOODE, WM. J., and HATT, PAUL K., *Methods in Social Research,* New York: McGraw-Hill Book Co., Inc., 1952, (386 pp.).

One of the best single texts for beginning social scientists, it treats in adequate detail all phases of research design and the employment of instruments used particularly in sociology and social psychology.

KAHL, RUSSELL, *Studies in Explanation,* New Jersey, Prentice-Hall, 1963, (363 pp.).

A reader in the philosophy of science, this small volume treats the basic problem of explanation through analysis of some outstanding writers from Ovid to Max Weber. A succinct yet lucid approach, with brief introductions by the editor.

KAPLAN, ABRAHAM, *The Conduct of Inquiry,* San Francisco, Chandler Publ. Co., 1964, (428 pp.).

Subtitled Methodology for Behavioral Science, Kaplan's excellent book brings together all the major issues involved in the scientific enterprise. His Section VII on Models is one of the best treatments to date; but all the ten sections are equally valuable.

KEMENY, JOHN G., *A Philosopher Looks at Science,* New York: D. Van Nostrand Co., Inc., 1959, (273 pp.).

A highly readable exposition of the basic problems linking science and philosophy.

LUNDBERG, GEORGE A., *Social Research,* New York: Longmans, Green & Co., Inc., 1942, (426 pp.).

A general text directed largely to social scientists. Though quite old now, it is notable as a prime exposition of operationalism and positivism in social science.

MADDEN, EDWARD H., *The Structure of Scientific Thought,* Boston, Houghton-Mifflin, 1960, (381 pp.).

This anthology is outstanding both for the selection of its arti-

cles and for the introductions to each of the seven sections prepared by the author. Add to this an excellent bibliography, and
Madden's book is one of the best introductions to the philosophy
of science extant today.

NAGEL, ERNEST, *The Structure of Science: Problems in the Logic
of Scientific Explanation,* New York: Harcourt, Brace & World,
Inc., 1961, (618 pp.).

A luminary in the field of the philosophy of science, Nagel
in this book updates the kinds of problems treated in Braithwaite and other such approaches. A very cogent treatment of
the epistemological problems of science.

NEURATH, OTTO, CARNAP, RUDOLF, and MORRIS, CHARLES (eds.),
International Encyclopedia of Unified Science, Chicago: University of Chicago Press, 1955, (2 vols.).

A landmark in the philosophical and methodological prob-
lems of empirical science. Typical of the highly erudite and sig-
nificant monographs included in this collection are those by
V. F. Lenzen, "Procedures of Empirical Science" (Vol. I, #5);
E. Nagel, "Principles of the Theory of Probability" (Vol. I, #6);
M. Cohen, "The Scientific Enterprise in Historical Perspective"
(Vol. II, #2); and C. G. Hempel, "Fundamentals of Concept
Formation in Empirical Science" (Vol. II, #7).

NEWMAN, JAMES R. (ed.), *What Is Science?* New York: Simon &
Schuster, Inc., 1955, (493 pp.).

An anthology by twelve eminent specialists representing the
major fields of science. Written for the layman, this book covers
in essential detail the interests, problems and methods of the
major fields of science.

NORTHROP, F. S. C., *The Logic of the Sciences and the Humanities,* New York: Meridian Books, 1959, (402 pp.).

Like Black or Cohen and Nagel, a very good general treatment
of the basic logical processes employed in scientific analysis.

PEARSON, KARL, *The Grammar of Science,* New York: Meridian
Books, 1957, (394 pp.).

A classic since the first edition in 1892, this book occupies a
distinguished place in the literature of science. Though obviously
outmoded in some sections, the discussion of basic principles

of scientific method has never been surpassed. A "must" reading for any student of the subject, though in no sense a handbook.
SELLITZ, CLAIRE *et al.*, *Research Methods in Social Relations*, New York: Dryden Press, 1959, (622 pp.).

One of the best books on research methods in the behavioral sciences, with special emphasis upon the problems and techniques involved in the measurement of prejudice.
TATON, RENE (Ed.), *History of Science*, New York, Basic Books, 1963-4-5, (3 vols., 1842 pp.).

Of the innumerable histories of science, this modern treatment is one of the best compromises between brevity and uninteresting detail. Especially valuable are the treatments of the social issues raised by the changing nature of science since its early beginnings.
THORNDIKE, LYNN, *A History of Magic and Experimental Science*, New York: Columbia University Press, 1929–1958, (8 vols.).

Unquestionably the most extensive treatment of the historical development of modern science. An excellent reference sourcebook.
TOWNSEND, JOHN C., *Introduction to Experimental Method*, New York: McGraw-Hill Book Co., Inc., 1953, (220 pp.).

An excellent short text in experimental method as employed in the behavioral sciences, particularly psychology.
WHITNEY, FREDERICK L., *The Elements of Research* (3rd. ed.), New York: Prentice-Hall, Inc., 1950, (539 pp.).

Generally a very adequate text, especially for advanced workers in educational research. More parsimonious yet more advanced than Good and Scates, it represents an effective compromise between simplicity and complexity, between minimal and maximal coverage of many problems of scientific method.
WILSON, E. BRIGHT, *An Introduction to Scientific Research*, New York: McGraw-Hill Book Co., Inc., 1952, (375 pp.).

One of the best books available which adequately treats the methodological and technical aspects of scientific research. Slanted toward the physical sciences, this is not for beginners, but not for professionals either. Requires advanced knowledge of mathematics.

Index

agreement, method of, 196 *ff*.
all-or-none fallacy, 97
analogy, imperfect, 98–100
analytic method, 205 *ff*.
  basic design in, 206–207
  comparative method, 210–211
  ex post facto design, 209–210
  fallacies in, 211–212, 234–235
  four-celled design, 208–210, 227–228
  function of, 205
  requirements for, 206
  role of variables, 212–216, 228–229
analytic study, 107–108
animism, 211
anthropomorphism, 211
a priori theory, 189
argument, sound and unsound, in logic, 47–51
assumptions, 34–37, 219–222, 243
astrology, *see* pseudo-science
authority, 25–26
  false, 92–93
  verification by, 238–242
axiom, 35–36

before-after design, 208–209
belief, 23, 24, 35
bibliography, 136–137
black-and-white fallacy, 96–97

capital punishment, 59–60
"card-stacking," 154–155
causation, 54, 107–108, 118, 215–216, 218
  coincidental, 187–188
  determination of, 180–186
  extrapolative, 183
  formal, 182
  function in science, 181
  functional, 184–186
  genetic, 182–183
  incidental, 187–188
  natural, 37–38

necessary, 188
notion of, 181
probabilistic, 184–185
probability theory of, 188–189
recognition of, 228
relation to theory, 190–192
significance of, 218–219
statistical, 218
teleological, 184–186, 234
types of, 186–190
certitude, 24
ceteris paribus reservation, 195
chance, 40–41, 235
common sense, 23, 110–111, 219–220
comparability, 194–196
comparative method, 210 *ff*.
concepts, 77–83, 116, 167, 168
  fallacious, 86–98
concomitant variation, 201–202, 216–217
consensus, 167, 235, 238–240
construct, 77–80, 112
controlled experiment, 205, 214–216
  frequency distribution, 215
  pair matching, 214
  precision matching, 214
crime, 64–65
cultural values, 8–9

data: area of, 125
  classification of, 172
  collection of, 154 *ff*.
  conclusiveness of, 126
  degree of generality of, 124–125
  delimitations of, 124–127
  evaluation of, 134–137
  frame of reference of, 218–222
  historical, 134–136, 140–143
  interpretation of, 224 *ff*.
  observation of, 154–163
  physical, 175
  presentation of, 171–176
  provided by literature, 132

253